T0206234

Optimizing Engineering Problems through Heuristic Techniques

Science, Technology, and Management Series

Series Editor:
J. Paulo Davim, Professor
Department of Mechanical Engineering, University of Aveiro, Portugal

This book series focuses on special volumes from conferences, workshops, and symposiums, as well as volumes on topics of current interested in all aspects of science, technology, and management. The series will discuss topics such as, mathematics, chemistry, physics, materials science, nanosciences, sustainability science, computational sciences, mechanical engineering, industrial engineering, manufacturing engineering, mechatronics engineering, electrical engineering, systems engineering, biomedical engineering, management sciences, economical science, human resource management, social sciences, engineering education, etc. The books will present principles, models techniques, methodologies, and applications of science, technology and management.

For more information about this series, please visit: https://www.crcpress.com/Science-Technology-and-Management/book-series/CRCSCITECMAN

Optimizing Engineering Problems through Heuristic Techniques

Kaushik Kumar, Divya Zindani, and
J. Paulo Davim

CRC Press
Taylor & Francis Group
Boca Raton London New York

CRC Press is an imprint of the
Taylor & Francis Group, an **informa** business

CRC Press
Taylor & Francis Group
6000 Broken Sound Parkway NW, Suite 300
Boca Raton, FL 33487-2742

First issued in paperback 2021

© 2016 by Taylor & Francis Group, LLC
CRC Press is an imprint of Taylor & Francis Group, an Informa business

No claim to original U.S. Government works

ISBN-13: 978-1-138-48536-5 (hbk)
ISBN-13: 978-1-03-217632-1 (pbk)
DOI: 10.1201/9781351049580

Publisher's Note

The publisher has gone to great lengths to ensure the quality of this reprint but points out that some imperfections in the original copies may be apparent.

Visit the Taylor & Francis Web site at
http://www.taylorandfrancis.com

and the CRC Press Web site at
http://www.crcpress.com

Contents

SECTION I   Introduction to Heuristic Optimization

SECTION II   Description of Heuristic Optimization Techniques

PART I   Evolutionary Techniques

PART II   Nature-Based Techniques

SECTION III   Application of Heuristic Techniques Toward Engineering Problems

Preface

The authors are pleased to present the book *Optimizing Engineering Problems through Heuristic Techniques* under the book series *Science, Technology, and Management*. The book title was chosen by looking at the present trend and noticing a book in this area, covering various popular and recent heuristic optimization techniques and its application to engineering problems to attain optimal solutions, would come in handy for various academicians, students, researchers, industrialists, and engineers.

Optimization is finding a solution or an alternative with the most cost effective or highest achievable performance under the given constraints, by maximizing desired factors and minimizing undesired ones. Optimization can be used in any field as it involves in formulating process or products in various forms. It is the "process of finding the best way of using the existing resources while taking into the account of all the factors that influences decisions in any experiment." The final product not only meets the requirements from an availability standpoint, but also from a practical mass production criteria.

There are two distinct types of optimization techniques: one traditional (statistical- and calculus-based), which is deterministic in nature, and the other heuristic, which is probabilistic in nature. The former has been in use for quite some time and has been successfully applied to many engineering problems. The heuristic technique is comparatively new and is gaining wide popularity due to certain properties which the traditional technique lacks. Due to complexity in engineering problems, an application engineer cannot afford to rely on a particular method and should know the advantages and limitations of various techniques, and therefore choose wisely the most efficient technique for the problem at hand. Heuristic optimization techniques are generally and presently being primarily utilized for non-engineering problems.

The book has **13 chapters** categorized into three parts, namely **Section I: Introduction to Heuristic Optimization Techniques, Section II: Description of Heuristic Optimization Techniques and Section III: Application of Heuristic Techniques towards Engineering Problems. Section I** contains **Chapter 1**, whereas **Section II** comprises **Two Parts. Part 1** has **Chapter 2** and **Chapter 3** describing the two most popular evolutionary techniques, namely **Genetic Algorithm** and **Particle Swarm Optimization. Part 2**, dedicated to **Nature-Based Techniques**, of this section has **Chapter 4** to **Chapter 7** describing four popular techniques, namely **Ant Colony Optimization, Bees Algorithm, Firefly Algorithm** and **Cuckoo Search Algorithm**, respectively. The last section, **Section III**, enlists **Chapter 8** to **Chapter 13**.

Section I, Chapter 1 introduces readers to the concept of heuristics and presents an overview of the same. Many real-life problems are modeled and solved for optimality through classical optimization techniques. One such class of optimization techniques is that of Heuristic search. Although heuristics do not guarantee optimality, they produce concrete results. Heuristics have been widely applied in various industries, such as business, statistics, environment, engineering, and sports.

Chapter 2, the first chapter of **Section II Part 1**, illuminates its readers with the fundamental concepts, mathematical models, and operators associated with genetic algorithm (GA). It is, no doubt, one of the most well-known and popular evolutionary algorithms. GA mimics the Darwinian theory of survival of the fittest in nature. The chapter also highlights improvements made in various components of GA, i.e., selection, mutation and crossover.

Particle swarm optimization (PSO) is discussed in the next chapter i.e., **Chapter 3**. PSO was proposed by Kennedy and Eberhart in 1995 and is a heuristic global optimization technique and now one of the most commonly employed. The present chapter delineates comprehensively an investigation into PSO and the advances made. The authors think this chapter would be beneficial for researchers involved directly or indirectly in the field of optimization.

Chapter 4, the first chapter of **Section II Part 2**, presents a brief overview of the structure of Ant Colony Optimization (ACO), its variants and the engineering applications. ACO has received considerable attention and has therefore emerged as one of the prominent Nature-Based Heuristic Optimization Techniques. ACO solves NP hard problems inspired by ant foraging behavior i.e., searching for food, the heuristics used by ants and the partial guidance of the other ants in indirect format. In this chapter, the components and the goals of ACO have also been depicted.

Chapter 5 provides an overview of the Bees Algorithm. The foraging behavior of honeybees is modeled by the Bees Algorithm and hence solves optimization problems. Exploitative neighborhood search in combination with the random explorative search is performed by this algorithm to solve optimization problems. The Bees Algorithm can be divided into four parts: tuning of parameter, initialization, the local search process, and at last the global search processes. In the present chapter, various improvements along with the application of the Bees Algorithm are discussed.

Chapter 6 presents a comprehensive outlook of firefly algorithm. The Firefly Optimization Algorithm has gained its stature from a so-called swarm intelligence. This algorithm has been applied to a number of domains including the field of engineering. The Firefly Optimization Algorithm has been able to successfully solve a variety of problems from different areas. Modified and hybrid variants of the Firefly Algorithm have been developed and hence its application scope has grown exponentially. Biological foundations of the Firefly Algorithm are also discussed in this chapter. The structure, characteristics and modified variants of firefly algorithms are discussed. Towards the end of the chapter, engineering applications to which firefly algorithms have been applied are discussed.

Chapter 7, the last chapter of **Part 2** as well as **Section II**, provides a brief overview of the Cuckoo Search Algorithm. Yang and Deb developed this in the year 2009 inspired by bird family. The present chapter also provides various applications of the optimization technique. From the chapter, it can be clearly observed that this algorithm has been used to address a wide range of engineering problems. The main objective of this chapter is to illuminate the readers with a definition of the Cuckoo Search Algorithm and also provide an outlook of the application areas it has addressed so far.

Section III, the section dedicated to solving engineering problems with heuristic techniques, starts with **Chapter 8**. The chapter describes the application of genetic algorithm to a non-traditional machining process i.e., ultrasonic machining process, which is one of the most extensively used non-traditional machining processes for the machining of non-conductive brittle materials such as glasses, carbides and bio-ceramics. The empirical models required for the optimization process were generated using the response surface methodology. Genetic algorithm has been applied to minimize the roughness for a hole surface. For optimizing the process parameters, different parameters considered were, namely, power rating, concentration of abrasive slurry and feed rate of the tool. As both the output parameters i.e., surface roughness and material removal rate are equally important, this becomes a multi-objective optimization.

The next chapter, **Chapter 9,** deals with the optimization problem for the electrical discharge machining process, another non-traditional machining technique. Setting optimal parameters, maximizing the material removal rate and minimizing the wear of the electrode tool, has been arrived at by employing the Particle Swarm Optimization technique (PSO). Once again, response surface methodology has been employed to arrive at the relationship between the inputs and outputs of the machining process, and the effectiveness of PSO algorithm has been demonstrated to address the optimization problem in an engineering domain.

In **Chapter 10**, the Ant Colony Optimization (ACO) technique has been employed to deal with the optimization problem in the multi-pass pocket milling process. Milling has been considered to be one of the oldest material removal processes that aids in removal of unwanted material through the use of rotating cutting tool. Setting optimal parameters, considering process parameters like speed of the spindle, depth of cut and feed rate, minimize surface roughness and machining time. The efficacy and suitability of the optimization technique have been demonstrated to address the optimization problem in the domain of a traditional machining process.

Following this trend, **Chapter 11** demonstrates the applicability of the Artificial Bee Colony Optimization algorithm, in order to determine the optimal combination of parameters for the Nd:YAG laser beam machining process by considering both the single- and multi-objective optimization of the responses. Nd:YAG laser beam machining process is one of the prominent non-conventional machining processes which has the potential ability to manufacture intricately shaped micro-products; however, identification of a suitable combination of parameters in order to achieve the desired machining performance is the key and the optimization technique serves it well.

Chapter 12 describes the application of the Firefly Algorithm to find an optimal solution for the electrochemical machining process. All the non-traditional machining processes, including electrochemical process, produce complex parts with great precision and are therefore time-consuming as well as expensive. Hence, it is necessary to select optimal parameters so that performance parameters such as heat affected zone (HAZ), radial overcut (ROC), and material removal rate (MRR) can be optimized. The Firefly Algorithm discussed, in this chapter was revealed to be robust and better in comparison to the results obtained by previous researchers.

Chapter 13, the final chapter of the book, illustrates the applicability of the Cuckoo Search Algorithm to predict surface roughness in the case of abrasive water jet machining. The Cuckoo Search Algorithm is one of the newest nature-based algorithms. Various models of prediction have been developed with different initial eggs, and analysis was carried out to investigate the best predicted value for surface roughness. The validity of the results has been established by employing the t-test, which ascertains applicability of the Cuckoo Algorithm for improving the performance of abrasive water jet machining. The results have revealed that the Cuckoo Algorithm is capable of optimizing process parameters that produce improved surface finish of the abrasive water jet machining process.

First and foremost we would like to thank God for allowing us to pursue our dreams. Almighty, without your support and blessings this work could not have been done. We would like to thank our ancestors, parents, and relatives for allowing us to follow our ambitions. Our families showed patience and tolerance while we took on yet another challenge that decreased the amount of time we get to spend together. They are our inspiration and motivation. We will be pleased if the readers of this book benefit from our efforts.

We would also need to thank all our well-wishers, colleagues, and friends. Their involvement in the development of this book cannot be overstated.

We owe a huge thanks to all of our technical reviewers and editorial advisory board members, our book development editor, and the team at CRC Press, for their work on this huge project. All of their efforts helped create this book. We couldn't have done it without their constant coordination and support.

Last, but definitely not least, we would like to thank everyone who took the time to help us during the process of writing this book.

Kaushik Kumar
Divya Zindani
J. Paulo Davim

Authors

Kaushik Kumar, B.Tech (Mechanical Engineering, REC (Now NIT), Warangal), MBA (Marketing, IGNOU) and Ph.D. (Engineering, Jadavpur University), is presently an Associate Professor in the Department of Mechanical Engineering, Birla Institute of Technology, Mesra, Ranchi, India. He has 18 years of teaching & research experience and over 11 years of industrial experience in a manufacturing unit of global repute. His areas of teaching and research interest are Conventional and Non-Conventional Quality Management Systems, Optimization, Non-Conventional machining, CAD/CAM, Rapid Prototyping and Composites. He has 9 Patents, 28 Books, 19 Edited Book Volumes, 43 Book Chapters, 141 International Journal, 21 International and 8 National Conference publications to his credit. He is Editor-in-Chief, Series Editor, Guest Editor, Editor, Editorial Board Member and Reviewer for International and National Journals. He has been felicitated with many awards and honors.

Divya Zindani, (B.E., Mechanical Engineering, Rajasthan Technical University, Kota), M.E. (Design of Mechanical Equipment, BIT Mesra), presently pursuing Ph.D. (National Institute of Technology, Silchar). He has over 2 years of industrial experience. His areas of interests are Optimization, Product and Process Design, CAD/CAM/CAE, Rapid prototyping and Material Selection. He has 1 Patent, 4 Books, 6 Edited Books, 18 Book Chapters, 2 SCI Journal, 7 Scopus Indexed International Journal and 4 International Conference publications to his credit.

J. Paulo Davim received his Ph.D. degree in Mechanical Engineering in 1997, M.Sc. degree in Mechanical Engineering (materials and manufacturing processes) in 1991, Mechanical Engineering degree (5 years) in 1986, from the University of Porto (FEUP), the Aggregate title (Full Habilitation) from the University of Coimbra in 2005 and the D.Sc. from London Metropolitan University in 2013. He is Senior Chartered Engineer by the Portuguese Institution of Engineers with an MBA and Specialist title in Engineering and Industrial Management. He is also Eur Ing by FEANI-Brussels and Fellow (FIET) by IET-London. Currently, he is Professor at the Department of Mechanical Engineering of the University of Aveiro, Portugal. He has more than 30 years of teaching and research experience in Manufacturing, Materials, Mechanical and Industrial Engineering, with special emphasis in Machining & Tribology. He has also interest in Management, Engineering Education and Higher Education for Sustainability. He has guided large numbers of postdoc, Ph.D. and master's students as well as has coordinated and participated in several financed research projects. He has received several scientific awards. He has worked as evaluator of projects for ERC European Research Council and other international research agencies as well as examiner of Ph.D. thesis for many universities in different countries. He is the Editor-in-Chief of several international journals, Guest Editor of journals, Books Editor, Book Series Editor and Scientific Advisory for many international journals and conferences. Presently, he is an Editorial Board member of 30 international journals and acts as reviewer for more than 100 prestigious Web of Science journals. In addition, he has also published as editor (and co-editor) more than 100 books and as author (and co-author) more than 10 books, 80 book chapters and 400 articles in journals and conferences (more than 250 articles in journals indexed in Web of Science core collection/h-index 52+/9000+ citations, SCOPUS/h-index 57+/11000+ citations, Google Scholar/h-index 74+/18000+).

Section I

Introduction to Heuristic Optimization

1 Optimization Using Heuristic Search

An Introduction

1.1 INTRODUCTION

Classical optimization techniques such as network-based methods, dynamic programming, non-linear programming, integer programming, linear programming, etc. can be used to model and optimally solve many real-life applications. These optimization techniques address different domains of research: operational research, scientific and engineering, scientific and computer science Sand management science. However, there are umpteen situations wherein the combinatorial nature of the problem makes it difficult to determine the optimal solution using the aforementioned classical optimization approaches. The time required from computational perspective is too large which is unrealistic to be acceptable in real-life applications. Furthermore the solution obtained may not be the optimal one i.e., global best and may be one of the local optima which may be relatively poor in comparison to the global best. Heuristic methods have been devised to overcome the aforementioned drawbacks and therefore aims to provide the user with a reasonably good solution.

There are certain cases wherein heuristics only seem to be a way forward to obtain concrete results. There has been wide range of application areas for heuristics such as business, economics, statistics, engineering, medicine and sports. Heuristics are now being adopted to solve wide range of complex problems that were very difficult to be solved earlier. The performance analysis of various heuristics can be adjudged through a number of measures.

"Heuristic" is a Greek word that means to discover and explore. Heuristics are referred to as approximate techniques. The major objective of heuristics lies in to construct an optimization model that is easily comprehendible and provides for good solutions in a reasonable computational time. There are number of combinatorial factors involved with such techniques such as statistics, computing, mathematical logic and human factors as such experience. Human experience in one of the crucial factors in designing a heuristic that can approach a solution faster and will be more relevant to the real-life situation.

The remainder of chapter is organized into following: the manner in which a real-world problem is approached is briefly discussed which is followed with brief discussion on some performance measures for the evaluation of a given method. Categorization of heuristics has been depicted next.

1.2   THE OPTIMIZATION PROBLEM

For the minimization problem, a general optimization model can be defined in the following form:

$$\left\{ \begin{array}{ll} \text{Minimize} & F(X) \\ \text{st} & X \in S, S \subseteq E \end{array} \right. \tag{1.1}$$

There are cases wherein it becomes difficult to solve Equation (1.1), mainly because of the following reasons:

i. E being the solution space can be finite or very large set which makes the problem as combinatorial optimization problem, or $E = R^n$ i.e., a continuous optimization problem or $E = N^n$ i.e., an integer optimization problem.

ii. X being the decision variable may be integer, binary, continuous or combination of any of these types.

iii. $F(X)$ being the objective function may not be continuous, linear or even convex and may be made up of several conflicting objectives.

iv. S being the feasibility set may not be convex and may be made of disconnected subsets.

v. The parameter values within definition of F and S can be probabilistic, estimated or even unknown.

The optimization problem falls into a discrete optimization problem if the solution set S is discrete and if it is continuous then the optimization problem is considered to be continuous optimization problem.

1.2.1   LOCAL VERSUS GLOBAL OPTIMA

Let $X \in S$ and the neighborhood of X may be represented by $N(X) \subset S$. $N(X)$ may be defined by a small area in the vicinity of X.

$\tilde{X}$ is a local minima or maxima with respect to its neighborhood if $F(\tilde{X}) \leq (\geq) F(X) \ \forall X \in N(X)$.

X^* is a global minima or maxima if $F(X^*) \leq (\geq) F(X) \forall X \in S$.

As for instance if all the neighborhoods is represented by ψ and set of all local minima or maxima is represented by Φ then global minima or maxima X^* can be defined as $X^* = \text{Arg Min} \{ F(X); X \in \psi(X) \}$ or $X^* = \text{Arg Min} \{ F(\tilde{X}); \tilde{X} \in \Phi \}$.

In short, global minima or maxima X^* is the local minima or maxima if it yields the best solution for the objective function under consideration.

Another mechanism is that of local search wherein $\tilde{X}$ is obtained from X in a given neighborhood $N(X)$. i.e., in other words $\tilde{X} = \text{Arg Min} \{ F(X); X \in N(X) \}$.

1.3   CATEGORIZATION OF OPTIMIZATION TECHNIQUES

There are two main categories wherein the optimization techniques falls into: exact algorithms and approximate or heuristic algorithms. The exact optimization algorithms guarantee optimal solution in a number of finite steps, whereas the other category involves heuristic which are set of rules developed through experience, mathematical logics and common sense. Heuristics have the potential ability to tackle the problems in a reasonable amount of computational time. However, the solution produced by such algorithms may not be optimal. Comparison of performances of such algorithms can be done using certain criteria and a discussion on this will be made in the subsequent chapters.

Although approximation and heuristic algorithms yield feasible solution, there exists some differences between heuristics and approximation algorithms. The difference lies that approximations guarantee quality of the solutions on the basis of worst-case scenarios. As for instance, the Christofides' algorithm that is used for solving the traveling salesman problem has a worst quality ratio of 1.5. Another example is that of next fit algorithm that has the worst quality ratio of 2. On the other hand, most heuristics have no similar mathematical bounds that can aid in adjudging their quality. However, research in this direction is underway to evaluate the quality of such non-optimal algorithms.

A possible approach to complex real-life problems are: (i) the objective should be to apply an exact methodology to the real-life complex problem, if this is not possible then step (ii) must be approached i.e., application of heuristic approach to an exact problem, if not possible then step (iii) must be approached: application of the exact method to the modified optimization problem and if this step is not approached then final step (iv) must be followed: application of heuristic approach to an approximated problem. The main idea lies in to maintain the characteristics of identified problem and then try to apply steps (i) and (ii).

The level of modification to the true problem must be considered carefully. A major modification may make it easier to solve the problem but the modified problem will have a very little resemblance to the originally identified problem. On the other hand, it will be tedious to approach a little modified problem.

Another plausible approach may be to incept with an easier version of optimization problem and then proceed to find a solution while keeping a check on the complex constraints. If the complex constraints are satisfied then there is no need to worry about the optimization problem under consideration. However, if any of the constraints are violated which is likely at the beginning of the search, then introduction of additional characteristic features is required. The process is iterated until it becomes impractical to solve the problem. The solution found in the previous stage then becomes the final solution of the optimization problem under consideration.

Hence modifications can be done at three different stages: input stage, algorithm stage and finally the output stage. Therefore it is one of the critical decisions as to when to consider the modifications. That said, it is always better to try for modifications either at the algorithm stage or the output stage. However, if by making slight changes to the initial identified problem can aid in solving the problem optimally, then such approach should be considered judiciously.

A less favorable result could be produced with reference to the interrelationship between the end user and the researcher. This could happen owing to the lack of understanding and lack of appreciation of the difficulties encountered while approaching a solution for the real-life complex problem. As a result of such outcomes the practitioners will distant themselves from the world of academics. However, on a positive note and owing to the better relationship between the universities and the outside world, the trend of distancing is diminishing. In such an environment the company gains an added advantage and the academician enriches their research portfolio. The enhanced research portfolio may benefit the faculty when they are adjudged for their excellence by the universities (2014).

1.4 REQUIREMENT OF HEURISTICS AND THEIR CHARACTERISTICS

As discussed in the aforementioned discussion that heuristics can only be used when there is impracticality with the employability of exact solutions which guarantee optimal solutions. This may arise either because of the excessive computational effort required or there is a potential risk of solution being trapped in local optimum.

Therefore in abovementioned circumstances, heuristics become virtually the only option to aid practitioners in finding reasonably acceptable solution. Some of the favorable reasons for promoting heuristics are as follows (Salhi, 2006): (i) heuristics aid the users to obtain solutions of large and combinatorial optimization problems, (ii) heuristics present a better understanding of the search progress through graphical representations, (iii) such algorithms are easy to code and implement, (iv) these algorithms are suitable for producing a number of feasible solutions and not a single one and therefore provides flexibility to the users to choose from more than one feasible solutions, (v) heuristics are easily accessible and adaptable to additional tasks or constraints and (vi) even personnel who have only superficial knowledge can well understand the process of optimization.

While designing a heuristic algorithm, there are certain characteristics that may be followed. Some of these are added for the generalization purpose and some are just the by-products of the attributes. Certain characteristics of heuristics have been discussed below (Salhi, 2006):

i. Effective and robust: The designed heuristic must be able to provide near optimal solution for the different cases under study.
ii. Flexible: The flexibility must be there to incorporate any modifications. Flexibility to modify any design step may aid in accommodating new ideas and concepts and the optimization problem can be approached with retention of benefits of the originally designed heuristics.
iii. Efficient: The time required needs to be acceptable and therefore must be efficient.
iv. Simple: Designed heuristic must be able to follow well-defined steps.

However, care must be taken that the heuristic doesn't gets trapped into local optimum. Metaheuristics are higher levels of heuristics that are devised to reduce the risk of the local searches and heuristics to being trapped into a poor local optimum.

1.5 PERFORMANCE MEASURES FOR HEURISTICS

Performance measures of heuristics can be measured through the solution quality, computational efforts, time complexity and space complexity. Below are mentioned five measures of checking solution quality of designed heuristics that can help in testing a given heuristic.

i. *Worst-case analysis*: An example that can show the weakness of the algorithm, which is usually referred to as pathological example, needs to be constructed. However, finding such an example is difficult especially in case of complex problems. One of the major drawbacks for carrying out such theoretically strong analysis is that the problem that is under study rarely represents the case for worst-case analysis. It is beneficial to understand well the problem under consideration to identify whether it truly resembles the example for worst-case analysis. Worst-case analysis provides for useful measures as it guarantees the performance of the algorithm that isn't far from the real-life example.

ii. *Lower bounds*: One way is to solve relaxed problem i.e., either LP relaxation problems wherein the difficult constraints are removed or to solve the Lagrangean relaxation problems. However, lower bound solutions must be tight so that the quality of the heuristic solution can be adjudged suitably and therefore presents the main difficulty. If this is not the case then users may draw misleading conclusions.

iii. *Empirical testing*: This is based on the best solutions obtained by the already existing heuristics on a set of published data. The designed heuristics can be compared using certain measures such as worst deviation, average solution and the number of best solutions, etc. Empirical testing is the most prominent simpler approaches and can be used when results from past researchers exist. Although the accuracy of the testing approach is guaranteed but it only provides for statistical evidence.

iv. *Probabilistic analysis*: The density function of the problem under consideration needs to be determined which allows for statistical measures such as worst behavior and average to be calculated.

v. *Benchmarking*: One of the obvious ways in which the performance of heuristics can be compared is to compare the designed heuristics with the already existing benchmarking solutions. This provides an advantage to the practitioners even if their designed heuristics doesn't fair with the benchmark solutions as they will be able to earn for the improvements. If the results obtained are good then this may instill self-belief and confidence in the user.

A good understanding of the heuristics is vital as inferior solutions result in a wrong signal to the user which the user can only comprehend if the basics on heuristics are right and suitably conceived. Better comprehension not only results in avoiding communication hick-ups but also helps in construction of friendly atmosphere in which modifications can be easily implemented during the course of design of heuristics. There are certain cases wherein the initial runs are not perceived by the users and the user will only be able to incept with their feedback only when positive results are found.

Time complexity is another measure of performance for heuristics. Time complexity of an algorithm is measured through $O(g(n))$ where the size of the problem is denoted by n. The problem can be solved within a reasonable time if $g(n)$ is a polynomial function. However, it may be difficult to solve if $g(n)$ is an exponential function. Such type of solutions are known as NP hard.

Space complexity is less referenced performance measure in comparison to time complexity. However, it is critically important to understand the manner in which the data is stored and retrieved. Smallest data storage capacity will not only aid in efficient data handling but it can also save a large amount of computing time as it can avoid calculation of unnecessary information. The problem may be encountered not only when computational time is large but also arise in case when the computer runs out of memory. A large amount of storage capacity may be demanded by the heuristic even during its initialization phase. Hence certain ways around the problems need to be identified.

Computational effort is measured through both the space as well as the time complexity. Large or small computing time is relative term and is defined by the nature of the problem and the availability of the resources for computing. The time for interfacing are usually ignored although it can constitute an important part of the total computing times. If carried out by professionals then this additional time could be taken as constant.

It is the importance of the problem that dictates the impact of computing effort. As for instance the algorithm needs to be quick if the problem that is addressed by it is required to be solved once or twice a day. However, if the problem needs to be solved once every month or year then lesser priority can be given to the CPU time. In such cases, attention can be given to the quality of solution. As for instance the problems associated with identification of locations for new facility, purchasing of expensive equipment and planning the schedule of work for the employees and so on doesn't cares for the computational time taken by the optimizer. However, the quality of solution is very critical to such investigations.

Computational time can be saved through an efficient computer code. This can be achieved through minimization of already computed partial or full information. Tracking of already computed information through the aid of efficient data structures is another way of saving computational time. Introduction of reduction tests that helps to minimize the testing of certain tests also plays a critical role in reduction of computational time. This also doesn't affect the quality of final solution.

1.6 CLASSIFICATION OF HEURISTICS

Heuristics can be classified into the following ways:

  i. Classical and modern
 ii. One solution and multiple solution at a time
iii. Fast and dirty and slow and powerful
 iv. Stochastic and deterministic

In the present book, following categorization of heuristics have been considered:

   i. Evolutionary techniques
   ii. Nature-based techniques
   iii. Logical search algorithms

1.7 CONCLUSION

Present chapter provides an overview of the heuristics and their usage in practice. Certain measures of performance and suitable characteristics of heuristics have also been depicted in the chapter that can aid the readers especially in designing of such techniques. A classification scheme as well as the categorization of the heuristics that will be used in the book have been presented towards the end of the chapter.

Section II

Description of Heuristic Optimization Techniques

PART I

EVOLUTIONARY TECHNIQUES

2 Genetic Algorithm

2.1 INTRODUCTION

Computational intelligence is one of the fastest growing fields together with evolutionary computation in optimization sciences. There are number of optimization algorithms to solve real-world complex problems. Such algorithms mimic mostly the biology surrounding the nature. Most of the evolutionary algorithms have a similar framework. They incept with a population of random solution. The suitability of the solution obtained is adjudged through a fitness function. Through a number of iterations the solution obtained at each step is improved and the best one is chosen. Next set of solutions are then generated through combination of achieved best solution and stochastic selections. There are several random components associated with an evolutionary algorithm that select and combine solutions in each population. Therefore in comparison to the deterministic algorithms the evolutionary algorithms are unreliable in finding suitable solutions. Same solutions are obtained at each and every step by deterministic algorithms. However, slower speed and possibility of getting stagnated at local solution are the major problems of deterministic algorithms when applied to large-scale problems.

Evolutionary algorithms are heuristics and stochastic. This means heuristic information is employed to search part of search space. These algorithms promise to search only selected regions of the solution space through finding best solution in each population and then use the generated solutions to improve other solutions. Evolutionary algorithms are now being used on large-scale applications and therefore has gained wider popularity and flexibility. Consideration of optimization problems as black boxes is another advantage associated with the evolutionary algorithms. Genetic algorithm (GA) is one of the first and well-known evolutionary algorithms. The present chapter therefore discusses and analyzes GA.

2.2 GENETIC ALGORITHM

GA is inspired by theory of biological evolution that was proposed by Darwin (Holland, 1992; Goldberg and Holland, 1988). Survival of the fittest is the main mechanism which is simulated in the GA. Fitter has the highest probability of survival in nature. They transfer their genes to the next generation. In due course of time, the genes that allow species to be adaptable to the environment become dominant and play a vital role in the survival of the species of next generations.

GA is *inspired* by the chromosomes and genes and therefore reflects a true representation of an optimization problem wherein chromosome is representative of a solution and each variable of the optimization problem is represented by a gene. As for instance, an optimization problem will have ten number of genes and chromosomes if it has ten variables. Selection, crossover and mutation are the three main operators that are employed by the GA to improve the solution or the chromosome. Following sub-sections depict on these steps and also the representation of the optimization problem and the initial population.

A chromosome is made from genes that represents the variable set of a given optimization problem. The first step to use GA is to formulate the problem and define the parameters in the form of a vector. Binary and continuous are the two variants of GA. Each gene is assigned two values in case of a binary GA, whereas continuous values are assigned in case of a continuous GA. Any continuous value having upper and lower bounds can be used in case of the continuous GA variant. A special case of binary GA is wherein there are more than two values to make a suitable choice. In such special cases, more memory i.e., bits must be allocated to the variables of the problem. As for instance if an optimization problem has two variables each of which can be assigned eight different values, then for each variable, there is a requirement of three genes each. Hence number of genes for variable with n discrete values will be $\log_2 n$. Genes can be used until they are fed into fitness function and result in a fitness value. GA is referred to as genetic programming if different parts of a computer program is employed for each gene.

Set of random genes incepts the GA process. Equation (2.1) is used in case of binary GA:

$$X_i = \begin{cases} 1 & r_i < 0.5 \\ 0 & \text{otherwise} \end{cases} \tag{2.1}$$

where i-th gene is represented by X_i and r_i is any random number between 0 and 1.

Equation (2.2) is used in case of continuous GA to randomly initialize the genes:

$$X_i = (ub_i - lb_i) * r_i + lb_i \tag{2.2}$$

The upper bound for the i-th gene is represented by ub_i and the lower bound by lb_i.

The main objective of the initial population phase is to have uniformly distributed random solutions for all the variables. This is because these will be used subsequently in the following operators.

Natural selection is simulated by the *selection* operator of GA. The chance of survival is proportionally increased to fitness in case of natural selection. The genes are propagated to be adapted by the subsequent generations after being selected.

The fitness values are normalized and mapped to the probability values by the roulette wheel. The upper and lower bound of roulette wheel are 1 and 0, respectively. One of the individuals will be selected by generating a random number within this interval. The chances for an individual to get selected is represented by the larger sectorial area occupied by the individual in the roulette wheel.

However, one pertinent question that may arise in the mind of readers is that why the poor individuals are not discarded. It is worth noting that even the individuals that have lower fitness value may also be able to mate and contribute toward subsequent generation production. However, this is dependent on other important factors such as competition, territory and environmental situations. An individual with poor fitness value may have chance to produce excellent features in conjunction with genes of other individuals. Hence by not discarding poor solutions, a chance is given to the poor individuals so that good features remain.

Since the range of values changes and is problem dependent, normalization of values is very important. One of the issues that surround the roulette wheel is that it fails in handling the negative values. Therefore the negative values must be mapped to positive ones through fitness scaling as negative values may impact during the cumulative sum process.

Some of the other selection operators (Genlin, 2004) besides roulette wheel are: steady-state reproduction (Syswerda, 1989), proportional selection (Grefenstette, 1989), fuzzy selection (Ishibuchi and Yamamoto, 2004), truncation selection (Blickle and Thiele, 1996), rank selection (Kumar, 2012), Boltzmann selection (Goldberg, 1990), linear rank selection (Grefenstette, 1989), fitness uniform selection (Hutter, 2002), local selection (Collins and Jefferson, 1991), steady-state selection (Syswerda, 1991) and tournament selection (Miller and Goldberg, 1995).

The natural selection process aids in selection of individuals for the crossover step and are treated as parents. This allows for gene exchange between individuals to produce new solutions. Literature suggests a number of different methods of crossover. The chromosome is divided into two or three pieces in case of the easiest of the methods (Shenoy et al., 2005). The genes between the chromosomes are then exchanged. This can be visualized in Figure 2.2 clearly.

The chromosomes of two parent solutions are swapped with each other in the single-point crossover and therefore there is one crossover point. However, in case of double-point crossover, there are two crossover points i.e., the chromosomes of the parent solutions swap between these points. The other techniques of crossover as mentioned in different literatures are: uniform crossover (Semenkin and Semenkina, 2012), three parents crossover (Tsutsui et al., 1999), cycle crossover (Smith and Holland, 1987), position-based crossover (Fonseca and Fleming, 1995), masked crossover (Louis and Rawlins, 1991), half-uniform crossover (Hu and Di Paolo, 2009), partially matched crossover (Bäck et al., 2018), order crossover (Davis, 1985), heuristic crossover (Fogel and Atma, 1990) and multi-point crossover (Eshelman et al., 1989).

The overall objective of the crossover step is to ensure that genes are exchanged and the children inherit the genes from the parent solutions. The main mechanism of exploration in GA is the crossover step. There can be crossover using random points and hence the GA is trying to check and search for different combinations of genes coming from parents. This step therefore aids in exploration of possible solutions without the introduction of new genes.

Probability of the crossover i.e., P_c is an important parameter in GA that identifies the probability of accepting a new child. This parameter is a solution in the interval 0 and 1. For each child a random number is generated in the interval [0,1]. The child is propagated to the subsequent generation if the random number generated is less than the probability of crossover. If this is not the case then parent is propagated. This is also true with the nature wherein all the offspring don't survive.

The main issue associated with the crossover is the lack of introduction of new genes. If all the solutions become poor, the crossover mechanism will not result in generation of different solutions. Hence to consider this issue, GA also considers the *mutation* operator.

Changes in the genes are randomly created through the aid of mutation phase. Probability of mutation i.e., P_m is a parameter that is used for every gene in the chromosome of child generated using the crossover phase. The parameter P_m is a number in the interval 0 and 1. A random number is generated for each gene for the new child. The gene is assigned a random number in the said interval with the upper and lower bounds if the random number is less than P_m.

There are numerous mutation techniques: uniform (Srinivas and Patnaik, 1994), Gaussian (Hinterding, 1995), supervised mutation (Oosthuizen, 1987), varying probability mutation (Ankenbrandt, 1991), power mutation (Deep and Thakur, 2007), non-uniform (Neubauer, 1997), shrink (Tsutsui and Fujimoto, 1993) and uniqueness mutation (Mauldin, 1984). Mutation is also the main mechanism of exploration for the GA method. The reason may be attributed to the fact that the mutation operator allows for random changes in the solution and hence allows it to move beyond the search space.

The genes in the original chromosomes are produced as a result of crossover and mutation step. There may be chances that all the parents are replaced by children depending on the probability of mutation. Hence there may be possibility of doing away with the good solutions. In order to take care of this issue, another operator known as Elitism (Ahn and Ramakrishna, 2003) is employed. A large number of research studies on GA have revealed the importance of this operator in GA.

A very simple mechanism underlies the basic operation of this operator. The chromosome consists of the best genes in the current population and propagates it to the subsequent generation without any changes. Hence the solutions are not damaged by the mutation and crossover process leading to the creation of new population. The ranking of individuals on the basis of their fitness value updates the list of the elites.

2.3   COMPETENT GENETIC ALGORITHM

It is very useful to employ innovations for explanation of working mechanisms of GA. However, innovations themselves are not understood well and therefore pose difficulty. There is a dire need of principled and mechanistic way of designing GA in order to address and successfully solve the difficult problems across a wide range of real-life complex problems. Competent GAs have been developed in last decades, and as a result of great strides, GAs are now able to solve hard problems quickly with higher accuracy and reliability. Competent GAs are able to solve difficult problems in a scalable fashion and hence are convenient from a computational standpoint. Furthermore, the burden on a user to differentiate between a good coding is eased. In case wherein GA can adapt itself to the problem, the burden on user eases as otherwise the GA would be required to adapt to the problem through appropriate coding and GA operators.

Some of the important lessons associated with design of competent GAs are discussed. The discussion is, however, restricted to selector combinative GAs and on the facets of competent GAs. Designing of competent selector combinative GAs can be decomposed into number of design steps using Holland's notion of a building block (BB) (Holland, 1975). Although the design decomposition has been delineated by Goldberg (2002), a brief review of the decomposition process is discussed subsequently.

It should be known that GAs process BBs. Working of GA through the process of decomposition and reassembly forms the originating root for the conceptualization of selector combinative GA. The well-adapted set of features known as *building blocks* were regarded as the components of the effective solution (Holland, 1975). The key conceptual framework involves implicit identification of BBs for achieving good solutions and recombination of the identified BBs to achieve solutions with very high performance.

It is very critical to understand problems with hard BBs. It is a usual standpoint of cross-fertilizing innovation that the BBs are hard to acquire for problems that are hard. This may be because of the associated complexity with the BBs. Furthermore it may be due to the fact that BBs are very hard to be identified and separated. The deceptive and misleading behavior of lower-order BBs is another reason for the same (Goldberg, 1987, 1989a; Goldberg et al., 1992b; Deb and Goldberg, 1994).

Another important consideration is to understanding of growth and time associated with the BBs. It is believed that the BBs exist in a kind of competitive market economy. As such steps must be taken in order to ensure that the best BBs grow and takeover as a dominant player in the market share of population. Also it is critical to understand that growth rate can neither be too fast or too slow. Setting of the crossover probability (P_c) and the selection pressure (s) such that Equation (2.3) is satisfied will aid in satisfying the growth in the market share:

$$P_c \leq \frac{1 - s^{-1}}{\epsilon} \tag{2.3}$$

where, ϵ is the probability of disruption of BB.

There are two other approaches to understand time. The basic tutorial associated with understanding time is beyond the scope of the book. However, for interested readers following examples have been delineated:

Selection-intensity models: Here the approaches in resemblance to the quantitative genetics (Bulmer, 1985) are used and modeling of the dynamics of the average fitness of the population is achieved.

Take over time models: Here the modeling of the dynamical aspects of the best individuals is achieved.

The convergence time t_c for a problem of size 1 and with all the BBs bearing equal importance or salience can be obtained using Equation (2.4) (Miller and Goldberg, 1995):

$$t_c = \frac{\pi}{2I} \sqrt{l} \tag{2.4}$$

where, I is the intensity of selection (Bulmer, 1985) and is dependent on the method of selection and the selection pressure. As for instance, for tournament selection, I can be obtained using Equation (2.5) (Blickle and Thiele, 1996):

$$I = \sqrt{2\big(\log(s)\big) - \log\big(\sqrt{4.14\log(s)}\big)} \qquad (2.5)$$

However, the convergence time will scale-up differently if the BBs have different salience. As for instance, the convergence time will be linear in case the BBs are scaled exponentially and can be calculated using Equation (2.6):

$$t_c = \frac{-\log 2}{\log\big(1 - I/\sqrt{3}\big)} l \qquad (2.6)$$

It is also quintessential to have a proper understanding on the supply and decision-making associated with the BBs. Ensuring adequate supply of raw BBs is one of the key role of the population. Larger number of complex BBs will be contained in a randomly generated population of increasing size. The population size, n, required to ensure that at least one copy of all the BBs remain can be obtained using Equation (2.7) (Goldberg et al., 2001):

$$n = \chi^k \log m + k\chi^k \log \chi \qquad (2.7)$$

where, m is the number of BBs, x is the number of alphabets in each of the BB and χ is the associated cardinality.

Decision-making among different BBs is another critical aspect besides ensuring the adequate supply of BBs. The decision-making is statistical in nature and the likelihood of making the best possible decision increases as the population size is increased. Therefore the population size required to not only ensure the adequacy of supply but also to ensure correct decision-making can be obtained using Equation (2.8) (Harik et al., 1999):

$$n = \frac{\sqrt{\pi}\sigma_{BB}}{2d} 2^k \sqrt{m} \log \alpha \qquad (2.8)$$

where, α is the probability of incorrectly deciding among the competing BBs, d/σ_{BB} is the signal-to-noise ratio. In brief the following components make up the population sizing model:

  i. Probabilistic safety factor: $\log \alpha$.
  ii. Subcomponent complexity which is quantified by m i.e., the number of BBs.
  iii. Competition complexity which is quantified by the total number of competing BBs i.e., 2^k.
  iv. The ease of decision-making which is quantified by d/σ_{BB}.

The population size scaling can be obtained using Equation (2.9) if there is exponential scaling of BBs (Rothlauf, 2006):

$$n = -c_o \frac{\sigma_{BB}}{d} 2^k m \log \alpha \qquad (2.9)$$

where, c_o is a constant and is drift effect dependent (Crow and Kimura, 1970; Asho and Muhlenbein, 1994).

One of the most important lessons in GA is the identification of BBs and their exchange. These two facets form the critical path to innovative success. It is a trend and observation that the first generation GA usually fail in their ability to promote reliably this exchange. The primary aspect of challenge associated with designing a competitive GA is the need to identify BBs as well as promote exchange among them. It has been revealed that although the recombination operators exhibit polynomial scalability for the case of simplified problem, they suffer from exponential scalability in case of boundedly difficult problems. The studies using facet wise modeling approach also reveal the inadequacies associated with the recombination operators in effective identification and exchange of BBs. A control map is yielded by mixing models suggesting regions of good performance related to GAs. Control maps can aid in identification of sweet spots for GA and hence help in parameter settings.

Research direction focused in designing effective GAs has led to the development of competent GAs and therefore in identification and exchange mechanisms for BBs. The developed competent GAs have the advantage of solving quickly the hard problems with greater reliability and accuracy. Hard problems are the problems that have very large sub-solutions which can't be decomposed into simpler sub-solutions or have umpteen minima or have high associated stochastic noise. The object is to develop an algorithm that can aid in solving the problems with bounded difficulties and exhibit polynomial scaling.

It is worth noting at this stage that there is a vast difference in the mechanics of competent GA. However, it is also true that there are invariant principles associated with innovative success. Messy GA markets the beginning of competent GA (Goldberg et al., 1989) which finally translated to give rise to fast messy GA. Thereafter a number of GA variants have been developed with the aid of different mechanism styles. Following discussion categorizes some of these approaches, however, a detailed discussion is beyond the scope of this book.

Probabilistic model building techniques: The prominent models include population-based incremental learning (Baluja, 1994), the compact GA (Harik et al., 1999), the Bayesian optimization algorithm (Pelikan et al., 2000), the hierarchical Bayesian optimization algorithm (Pelikan and Goldberg, 2001), etc.

Linkage adaptation techniques: The prominent examples include linkage learning GA.

Perturbation technique: Messy GA (Goldberg et al., 1989), fast messy GA (Goldberg et al., 1989), linkage identification by nonlinearity check (Munetomo and Goldberg, 1999), the dependency structure matrix driven GA (Yu et al., 2003).

2.4 IMPROVEMENTS IN GENETIC ALGORITHMS

In the previous section, discussion was made on competent GAs. The competent GAs have shown to solve successfully the hard problems and have yielded promising results. However, competent GAs only solve l-variable search problems, wherein $O(l^2)$ number of function evaluations are only required. Such problems are referred to have subquadratic number of function evaluations. The competent GAs have addressed the challenges associated with the first generation GAs and have rendered the intractable to tractable. But it can be daunting and tedious task to compute and evaluate subquadratic number of functions. Single evaluation may take long hours if the fitness function evaluation involves complex simulation or computing. Even the subquadratic number of function evaluations for such cases is very high. As for instance, half a months' time would be required to solve a 20-bit search problem given the fact that the evaluation of fitness function takes at least 1 h. The role of efficiency enhancement technique becomes critical in such cases. Furthermore, in order to make an approach really effective for a particular problem, GA needs to be integrated with problem-specific methods. There are numerous literature that have been discussed and investigated on the enhancement of GAs. The four major categories of GA enhancement have been discussed next with suitable references so that interested readers may connect as and when required.

Evaluation relaxation: Here the less accurate but inexpensive computationally fitness estimate replaces the computationally expensive and accurate fitness evaluation. The less accurate and low-cost fitness estimate can either be exogenous or endogenous. Surrogate fitness function is a case of exogenous fitness evaluation where the development of fitness estimate takes place through external means. Fitness inheritance is the case associated with endogenous function estimate wherein the fitness evaluations are done internally and is based on parental fitness (Smith et al., 1995).

Evaluation relaxation technique dates back to early and has built up on the empirical work in image registration by Grefenstette and Ftzpatrick (1985). Using the technique, significant speeds were achieved as the random sampling of the image pixels were reduced greatly. Since then, the technique occupied center stage and was employed to address complex optimization problems across different disciplines such as warehouse scheduling at Coors Brewery (Watson et al., 1999) and structural engineering (Barthelemy and Haftka, 1993).

Design theories have been developed to evaluate the effect on population sizing and convergence time that have progressed the early empirical studies on relaxation techniques. These developments have resulted in optimizing speed-ups in approximate functions.

Hybridization: It is one of the effective ways of enhancing the effectiveness and performance of GAs. Coupling of GAs with the local search techniques and incorporation of domain-specific knowledge is the most common hybridization technique. Incorporation of local search operator into GA is another common form of hybridization technique. The hybridization process aids in production of stronger results in comparison to the results that can be achieved using individual approaches. However, increased computational effort is one of the limitations associated with the hybridization techniques. Some of the examples in which case one can refer the process as hybridization of GAs are as follows:

  i. Repairing of infeasible solutions into legal ones.
  ii. Incorporation of experience of past attempts into the GA process.
  iii. Initialization of GA population
  iv. Development of specialized heuristic operators with combinative effects
  v. Decomposition of large problems into smaller sub-problems heuristically.

Significant successes with hybridization approaches have been revealed with the difficult real-world application areas. A small number of real-world examples addressed using hybridized GA have been mentioned below:

  i. Machine scheduling (Sastry et al., 2005)
  ii. Sports scheduling (Costa, 1995)
  iii. Warehouse scheduling (Watson et al., 1999)
  iv. Nurse rostering (Burke et al., 2001)
  v. Electric power systems such as maintenance schedule for thermal generator (Burke and Smith, 2000) and maintenance scheduling for electricity transmission network unit commitment problem
  vi. University timetabling such as timetabling for courses (Paechter et al., 1995) and timetabling for examinations (Burke et al., 2001).

Theoretical efforts have been scarce that underpins the hybridization of GA. Some efforts in the past have been made to address the modeling issues of GAs, to study the effect of sampling and search space and so on.

Parallelization: The GAs are run on multiple processors and there is distribution of computational resources among these processors. There are number of parallelization approaches such as simple master slave GA, a fine-grained architecture, a coarse-grained architecture or a hierarchical architecture. The key objective is to speed up the GA process by employing several processors that take up the computational loads.

Time continuation: A solution possessing high quality is achieved through the capabilities associated with recombination and mutation. The solution of high quality is obtained within the constraint of computational resource. A tradeoff between the small solution with multiple convergence epochs and the large population with single convergence epochs is obtained using the concept of time relaxation or continuation.

2.5 CONCLUSION

The present chapter delineated the main mechanism of GA i.e., mutation, recombination and initialization. The most widely used approaches for the main mechanisms were discussed in detail. The first generation of GA can solve problems with discrete variables and therefore competent GAs were developed. These developments have been depicted in detail in the present chapter. Different enhancements technique in improving the competent GAs have also been delineated.

3 Particle Swarm Optimization Algorithm

3.1 INTRODUCTION

Swarm intelligence falls under the realm of evolutionary computation. It researches the collective behavior of self-organized and decentralized systems irrespective of whether the systems are natural or artificial. Simple agents or boids interact locally with one another as well as the environment in swarm intelligence framework. Nature is the main source of inspiration for such intelligence techniques (Kothari et al., 2011). Simple and multiple rules are followed by the agents in swarm intelligence framework. There is no centralized structure for controlling the behavior of the agents in such frameworks. The behavior of agent in the framework are real and random to a certain degree, however, intelligent behavior at global scales emerge owing to the local interactions. This global behavior is unknown to the individual agents in the swarm intelligence framework. Some of the prominent examples of swarm intelligence include fish schooling, bacterial growth, animal herding and ant colonies.

An optimization algorithm based on bird flocking was proposed by Kennedy and Eberhart (Kennedy, 1995) and is referred to as particle swarm optimization (PSO). Some of the other intelligent optimization algorithms are differential evolution (Storn and Price, 1997), bacterial foraging optimization (Müller et al., 2000), artificial bee colony (Karaboga and Basturk, 2007a), glowworm swarm optimization (Krishnanand and Ghose, 2005) and bat algorithm (Yang, 2010a).

The present chapter focusses on PSO. Some of the studies on advancement of PSO have been presented. Various applications of PSO have also been depicted. Finally the chapter concludes with the conclusion that summarizes the improvements and the potential research directions.

3.2 BASICS OF PARTICLE SWARM OPTIMIZATION APPROACH

One of the key features of swarm intelligence is self-organization. It is a feature wherein due to the local interactions between the disordered components of the system, the global coordination or the order arises. The process is spontaneous and is not controlled by any inside or outside agent. The three basic ingredients of self-organization as identified by Bonabeau et al. (1999) are as follows:

i. *Multiple interactions*: Information from the neighbor agents is utilized by the agents in the swarm and therefore spread across the network.

ii. *Balance of exploration and exploitation*: A valuable mean approach of creativity is provided through a suitable means by the swarm intelligence algorithms.

iii. *Strong dynamical nonlinearity*: Convenient structures can be created from the positive feedback, while on the other hand the positive feedback also balances the negative feedback. This ultimately aids in stabilizing the collective pattern.

Besides the above features, five major principles identified by Milonas (Karaboga et al., 2014) to be satisfied by the swarm intelligence framework are: adaptability, stability, diverse response, quality principle and proximity principle. In accordance with the proximity principle the swarm intelligence must be able to do simple space and time computations. As a part of quality principle, the swarm must be able to respond to the quality factors in the environment. The swarm is also required not to commit its activities along excessively narrow channels as a part to fulfill the diverse response principle. In accordance with the adaptability principle, the swarm should be able to change their behavior as and when deemed suitable in accordance with the computational price. Furthermore, to fulfill the stability principle, the swarm must ensure so as not to change its mode of behavior every time there occurs change in the environment.

3.2.1 STRUCTURE OF STANDARD PSO

Swarm of particles are employed by PSO to perform the search operation. These swarm of particles update for every iteration. Each particle moves in the direction to the previous best position as well as the global best position in order to seek the optimal solution. The previous best i.e., pbest and the global best i.e., gbest are given by the following equation:

$$\text{pbest}(i,t) = \arg\min_{k=1...t}\left[f\left(P_i(k)\right)\right],$$

$$\text{gbest}(t) = \arg\min_{\substack{i=1...N_p \\ k=1...t}}\left[f\left(P_i(k)\right)\right] \quad i \in \{1,2,...,N_p\} \tag{3.1}$$

The particle index is represented by i, total of number of particles by N_p, fitness function is denoted by f, current iteration number by t and the position by P. Velocity V and Position P are updated in accordance with the Equations (3.2) and (3.3), respectively:

$$V_i(t+1) = \omega V_i(t) + c_1 r_1\left(\text{pbest}(i,t) - P_i(t)\right) + c_2 r_2\left(\text{gbest}(t) - P_i(t)\right) \tag{3.2}$$

$$P_i(t+1) = P_i(t) + V_i(t+1) \tag{3.3}$$

where, ω is referred to as inertia weights that is employed to balance the local exploitation and global exploration, r_1 and r_2 are the uniformly distributed random variables and are in the interval ranging 0 and 1, c_1 and c_2 are known as acceleration coefficients and are positive constants.

It is common practice to set up upper limit for the velocity parameter. To restrict the particles flying out of the search space, velocity clamping has been used (Shahzad, 2014). Constriction coefficient is another method that was proposed by Clerc and Kennedy (2002).

Inertia is represented in the first part of Equation (3.2) and provides the necessary momentum for the particles to roam across the search space. The second part of the Equation (3.2) represents the cognitive component and is significant of individual thinking of particle. This component is a motivational factor for the particles to progress toward their own best position. Cooperation component is the third part of the Equation (3.2) and reflects the collaborative efforts of the particles. This component aids the particle to search for the global optimal solution (Zhang et al., 2014).

Position and velocities are adjusted at each time step, and the optimization function is then evaluated for the new coordinates. The particle stores the coordinates in the vector $p_{\text{best id}}$ as and when the particle discovers a pattern that is better than the previously identified one. The difference between the current individual point and the best point identified by a particular agent is added to the current velocity stochastically and therefore the trajectory of the particle as such is caused to oscillate around the point. Furthermore, each particle is defined within the realm of topological neighborhood that comprises the particle itself and other particles in the population. Also the particle velocity gets updated through the addition of weighted difference between the global best and neighborhood best to its current velocity. This addition is also stochastic and hence the velocity is adjusted for the next time step.

3.2.2 SOME DEFINITIONS

Particle (X): This is candidate solution and is represented by d dimensional vector. The dimension of vector is defined by the number of optimized parameters. Particle at any time t can be depicted as $X_i(t) = [X_{i1}(t), X_{i2}(t),...,X_{id}(t)]$, where the optimized parameters are represented by X's and $X_{id}(t)$ reflects the position of ith particle w.r.t. to the value of the dth optimized parameter in the ith candidate solution.

Population X(t): The set of particles is reflected in population and is represented by $X(t) = [X_1(t), X_2(t)...X_n(t)]$.

Swarm: The disorganized population of moving particles is represented by swarm. In a swarm the particles tend to cluster with one another wherein each particle moves in a random direction.

Particle velocity V(t): The velocity of moving particles is represented by d dimensional vector. The velocity of a particle at any time t can be obtained using Equation (3.2). It is represented by $V_i(t) = [V_{i1}(t), V_{i2}(t)...V_{id}(t)]$ where the velocity of the ith particle with respect to the dth dimension. The value of $V_{id}(t)$ fluctuates between the range $-V_{\text{min}}$ and $-V_{\text{max}}$ and is therefore referred to as velocity clamping.

Inertia weight (w): The exploitation and exploration of the search space are controlled by the inertia weight. It dynamically adjusts the velocity. The effect on current velocities of the previous velocity is controlled using the inertia weight. A compromise between the global and local exploration abilities of the swarm is exhibited. Global exploration is facilitated through a large inertia weight wherein the local exploration is facilitated by a small weight. Therefore the inertia weight must be chosen carefully so as to provide a balance between the local and global exploration space. A proper balance between the two will result in yielding better solution. It is usually a better perspective to incept with a large inertia weight to provide a better global exploration and then decrease it to obtain a more refined solution.

Ability to search nonlinearly is one of the requirement often required by many search algorithms. Statistical features may be derived from the results obtained which will ultimately aid in understanding the PSO. This will ease the calculation of proper inertia weights for the next iteration. The inertia weight decreases linearly in accordance with the following equation:

$$w = w_{max} - \frac{w_{max} - w_{min}}{iter_{max}} \times iter \qquad (3.4)$$

where, w_{max} and w_{min} are the maximum and minimum values of inertia weights, the current iteration represented by iter and maximum number of iterations by $iter_{max}$.

Social and cognitive parameter: c_1 and c_2 represent the cognitive and social parameters. Each particle in PSO keeps track of its coordinates in the problem space and is associated with the best solution achieved so far. The best solution is referred to as particle best p_{best}. Another coordinate tracked is the overall best value of the particle and is represented by g_{best}. PSO aims to modify the values of particle position such that p_{best} and g_{best} are achieved. Constants c_1 and c_2 represent stochastic acceleration term that tends to pull a particle toward its p_{best} and g_{best}. Lower values of these constants causes the particle to move away from the target regions whereas abrupt movements are signified by the higher values.

It has been revealed that values of these constants if closer to 2 then good results are obtained usually. Furthermore, fast global convergence is achieved through this value. There is no significant changes in the rate of convergence with increasing value of these constants. Small local neighborhood aids in avoidance of local minima, however, faster convergence is obtained through larger global neighborhood.

3.3 PSO ALGORITHM

The steps involved in a PSO algorithm have been discussed below:

Initialization: The population of random particles is initialized wherein each of the particles have random velocity and position. The lower and upper limits for the decision variables are set to confine the search space of the solution. The initialized population of particles is such that the velocity as well as the position fall into the range of variables assigned and satisfies the constraints. A population size ranging 20–50 is more common in PSO algorithm.

The fitness of each particle is obtained in terms of pareto-dominance. The non-dominated solutions are recorded and are achieved. The memory of each individual is initialized and is used for the storage of personal best position. The global best position is chosen from the archive.

Velocity update: The velocity of each particle is updated in accordance with Equation (3.2).

Position updating: The position of particles are updated between successive iterations in accordance with Equation (3.3).

The feasibility of all the generated solutions are ensured through a check on all the imposed constraints. If in case any of the inequality constraint is violated by any element, then the position of the individual is fixed to its maximum or minimum operating point. Archive is also updated that stores the non-dominated solution.

Memory update: The particle's best position as well as the global best solutions are updated using the following equations.

$$\left. \begin{aligned} p_{\text{best}}(t+1) &= p(t+1)\, \text{if}\, f\big[p(t+1)\big] < f\big[p_{\text{best}}(t)\big] \\ g_{\text{best}}(t+1) &= p(t+1)\, \text{if}\, f\big[p(t+1)\big] < f\big[g_{\text{best}}(t)\big] \end{aligned} \right\} \tag{3.5}$$

where, $f(X)$ is the objective function that requires to be minimized.

The fitness evaluation of particles are compared with particles p_{best}. If current value is better than $p_{\text{best}}(t)$, then $p_{\text{best}}(t+1)$ is set as the new current value for subsequent iteration in the d dimensional space. The fitness evaluation is compared with the population's overall previous best. If the current global position $g_{\text{best}}(t+1)$ is better than $g_{\text{best}}(t)$ then the global best is set to $g_{\text{best}}(t+1)$.

Examination of termination criteria: The algorithm repeats the aforementioned steps until and unless a sufficient good fitness value is achieved or maximum number of iterations have been achieved. The algorithm, on termination, will generate the output points $g_{\text{best}}(t)$ and hence $f(g_{\text{best}}(t))$.

The optimal parameters that have been considered usually to yield optimal solutions are as follows: population size considered is 50, number of iterations as 100, c_1 and c_2 are set to 2, inertia weight w can range between 1.4 and 0.4.

3.4 SOME MODIFIED PSO ALGORITHMS

3.4.1 QUANTUM-BEHAVED PSO

The concept of quantum-behaved PSO (QPSO) stemmed from quantum mechanics. A modified QPSO was proposed by Jau et al. (2013) that aided in elimination of the associated drawbacks of basic PSO. The proposed algorithm employed a high-breakdown regression estimator as well as least-trimmed square method. QPSO with differential mutation operator was employed by Jamalipour et al. (2013) for optimization of WWER-1000 core fuel management. It was revealed that QPSO-Differential mutations (QPSO-DMs) performs better than the basic PSO algorithm. QPSO was used by Bagheri et al. (2014) for foreign exchange market. An improved QPSO algorithm was proposed by Tang et al. (2014) for continuous nonlinear large-scale problems which was based on memory mechanism and memetic algorithm. The memetic algorithm aided the particles to gain some experience through the local search phase and then utilize this experience for the subsequent evolutionary process. On the other hand the memory mechanism led to the introduction of bird kingdom and therefore improving the global search ability of the QPSO algorithm. A new hybrid approach encompassing QPSO and simplex algorithms was proposed by Davoodi et al. (2014) wherein QPSO was the main optimizer and simplex algorithm was used to fine-tune the solution obtained from QPSO. Artificial fish swarm algorithm was integrated with QPSO by Yumin and Li (2014). Jia et al. (2014) proposed an enhanced approach wherein QPSO was based on genetic algorithm. Through the enhance approach, synchronous optimization of sensor array and classifier was achieved. An improved QPSO metaheuristics algorithm was proposed by Gholizadeh and Moghadas (2014) to be employed for performance-based optimum design process.

3.4.2 CHAOTIC PSO

Chaos theory has been integrated with PSO in order to improve the overall performance of the standard PSO. The integrated version is known as chaotic PSO (CPSO). Chaotic maps were introduced into catfish swarm optimization which ultimately resulted in increased search capability (Chuang et al., 2011). An adaptive PSO was proposed by Zhang and Wu (2011), which was ultimately used for the development of hybrid crop classifier. A chaotic embedded PSO was proposed by Dai et al. (2012) and employed for the estimation of wavelet parameters. The chaotic variables were embedded into standard PSO and the parameters were adjusted nonlinearly and adaptively. A novel algorithm based on CPSO and gradient method known as chaotic particle swarm fuzzy clustering was proposed by Li et al. (2012). The proposed algorithm combined the iterative chaotic map with the adaptive inertia weight factor and ultimately with infinite collapses based on local search. The chaotic particle swarm fuzzy clustering exploited the searching capability of fuzzy c-means and therefore avoided the major limitation of standard PSO getting stuck into local optima. The convergence of the novel algorithm was steadfast through the adoption of gradient operator. A novel support vector regression machine was proposed by Wu et al. (2013) and was utilized to estimate the unknown parameters associated with CPSO. A fitness scaling adaptive CPSO was proposed by Zhang et al. (2013) and was used for planning of path for an unmanned combat aerial vehicle. The robustness of the proposed algorithm was justified and it was revealed that the proposed algorithm optimized the problem in lesser time as compared to those obtained with genetic algorithm, simulated annealing and chaotic ABC. K2 algorithm was applied with CPSO to Bayesian structure learning (Zhang et al., 2013). Optimization of municipal waste collection in Geographic Information Systems (GIS)-based environment was done using CPSO by Son (Le Hoang, 2014). A novel hybrid model combining artificial neural network and CPSO was proposed by Lu et al. (2014) which improved the forecast accuracy of standard PSO. Classical PSO was combined with a chaotic mechanism, a self-adaptive mutation scheme and time-variant acceleration coefficients (Zeng and Sun, 2014). This eliminated the premature convergence and aided in improvising the quality of the solution. A different chaotic system was proposed by Pluhacek et al. (2014) based on pseudorandom number generators. This was then applied for velocity calculation in the classical PSO algorithm.

3.4.3 TIME VARYING ACCELERATION COEFFICIENT-BASED PSO

The performance of classical PSO was also improved with time varying acceleration coefficient and was referred to as PSOTVAC. A modified PSO with time varying accelerator coefficients was proposed to take care of the linear automation strategy and thereby giving rise to PSOTVAC in which a predefined velocity index aided in adjusting the cognitive and social factors. PSOTVAC has been employed to address the economic dispatch problem (Chaturvedi et al., 2009). TVAC was employed efficiently that controlled local as well as global search and hence was successful in avoiding the premature convergence. An optimal congestion management was approached by Boonyaridachochai et al. (2010) for deregulated electricity market. The redispatch cost was determined to be minimum with effective implementation of PSOTVAC. A comparative analysis between PSO and self-organizing hierarchical PSO with time varying acceleration coefficient was demonstrated by Sun et al. (2011) for data clustering application. It was revealed that the self-organizing PSO had better performance in comparison to the classical PSO approach. Furthermore, it was revealed that PSO algorithm performed better in case of large-scale and high dimensional data. An efficient approach for economic load dispatch problems was addressed by Abedinia et al. (2014) using the PSO with time varying acceleration coefficient. A realistic look to the problem was provided through constraints as transmission loss, ramp rate limit, prohibited operating zone, nonlinear cost functions and generation limitations. An iteration PSO with time varying acceleration coefficient was employed for solving economic dispatch problems and a good convergence property was revealed by the proposed heuristic algorithm (Mohammadi-Ivatloo et al., 2012). A time varying acceleration coefficient PSO was employed by Mohammadi-Ivatloo et al. (2013) to solve combine heat and power economic dispatch problem. The solution quality of original PSO was improved through adaptively varying the acceleration coefficients in PSO algorithm. A binary PSO with time varying acceleration coefficients was proposed by Pookpunt and Ongsakul (2013) and solved the problem associated with the optimal placement of wind turbines within a wind farm. The objective was to maximize the power output with minimum investment. A hybrid PSO with time varying acceleration coefficient integrated with bacteria foraging algorithm was proposed by Abedinia et al. (2013) to solve complex economic dispatch problem. A modified PSO with time varying acceleration coefficient was presented to address the economic load dispatch problem by Abdullah et al. A new best neighbor particle was employed to improve the quality of the solution of the classical PSO algorithm. A binary PSO with time varying acceleration coefficient and a chaotic binary PSO was presented by Zhang et al. (2015). These novel PSO algorithms were then used to solve the multidimensional knapsack problem. The proposed novel algorithms were found to be better to other methods in terms of mean absolute deviation, success rate, least error and standard deviation.

3.4.4 SIMPLIFIED PSO

Swarm were divided into three categories: ordinary particles, better particles and the worst particles by Chen (2010). The divide was done in accordance with the fitness value and three types of swarms evolved in accordance with the simplified PSO algorithms. Simplification of PSO was done by Pedersen and Chipperfield (2010) and the adaptability of the classical PSO was improvised. The behavior parameters were tuned using an overlaid metaoptimizer. The modification were incorporated in classical PSO and the version was referred to as many optimizing liaisons, and it was revealed through experimentations that the new PSO algorithm panned out well in comparison to the classical PSO version. A simplified PSO was proposed by dos Santos et al. (2012) and saving in computational time was revealed with better performance characteristics. Design and performance analysis of proportional-integral device was presented by Panda et al. (2012) using many optimizing liaisons PSO and employed it for an automatic voltage regulator system. A simplified PSO was proposed to address proportional-integral proportional derivative by Vastrakar and Padhy (2013). A parameter-free simplified PSO was proposed by Yeh (2013) and was used to adjust the weights in artificial neural networks (ANNs).

3.5   BENEFITS OF PSO ALGORITHM

PSO algorithm has the following advantages:

i. It can handle stochastic nature of objective function.
ii. It has the potential ability to handle very large number of operating processors and hence the capability to escape the local minima.
iii. Simple mathematical functions as well as logic operations are used and therefore easy to implement.
iv. It is a derivative-free algorithm.
v. It doesn't require initial good solution to guarantee its convergence.
vi. It can be easily integrated with the other optimization techniques.
vii. It requires lesser parameters to be adjusted.
viii. It can be used for discrete as well as continuous or discontinuous variables and objective functions.

3.6   APPLICATIONS OF PSO

There are few applications of PSO that is specific to mechanical engineering domain. Implicit relationship between mechanical properties and the composition of as-cast Mg-Li-Al alloys was established by Ming et al. (2012). A momentum back-propagation neural network with hidden layer was employed for revealing the relationship. A procedure combining finite element analysis (FEM) and PSO was proposed by Chen et al. (2013) and was used for reliability-based optimum design of the composite structure. A good stability of the proposed method was revealed and the method was observed to be efficient in dealing with the probabilistic nature of composite design. PSO technique was employed by Mohan et al. (2013) to aid frequency response function in detection and quantification of surface damage. A better accuracy was revealed with the proposed methodology due to the fact that the data comprised of natural frequencies as well as mode shape. A surrogate-based PSO algorithm was applied by Chen et al. (2013) and employed it for reliability-based robust design of pressure vessels. Maximization of performance factor was solved considering the following design variables: the winding orientation, drop-off region size and thickness of the metal liner. Tsia-Wu failure criterion was used to construct the strength constraints of metal liners and composite layers. A methodology for identification of parameter values of Barcelona basic model was presented by Zhang et al. (2013). The difference between the measured and computed values was minimized using a novel parallel PSO algorithm. Wang et al. addressed the design problem of tubular permanent magnet linear synchronous motor through a novel PSO underpinned by decomposition-based multi-objective differential evolution. PSO algorithm with finite element method was utilized to model a unique column test (Lazrag et al., 2013). This aided in identification of all the associated hydraulic parameters of sand. Several tensiometers were installed in different positions along the column and inverse analysis was performed using the proposed methodology. A simulation-based PSO algorithm was employed by Lee et al. (2014) to accurately define the best screw position and number of locking compression plates for a fractured femur. This aided in achieving the acceptable fixation stability. Geometry of slotted micro-electromechanical system (MEMS) was optimized using PSO (Lake et al., 2013) which resulted in reduced energy dissipation from thermoelastic dissipation. Fundamental physics was used in tandem to PSO optimization technique to overcome the complications arising out of multiphysical problems. A hybrid optimization technique combining finite element analysis (FEA) PSO and continuous genetic algorithm was adopted by Vosoughi and Gerist (2014) to detect damage in case of laminated composite beams. The integrated framework aided in identification of useful information to define the range of solution and hence selection of suitable geometric parameters. A PSO method for determination of heat transfer coefficients to be employed for finite element modeling of medium voltage switch gear was proposed by Kitak et al. (2014). PSO algorithm was applied by Kalatehjari et al. (2014) for determination of critical slip surface of soil slopes.

3.7 CONCLUSION

Present chapter delineated basic conceptual framework of PSO. PSO reveals a close adherence to the five principles of swarm intelligence (Karaboga et al., 2014). PSO algorithm has some drawbacks such as higher complexity associated with computation, sensitivity to parameters, slow convergence and so forth. One of the potential reason is that the PSO algorithm doesn't utilize the crossover operator as employed in differential evolution and genetic algorithm. Furthermore, PSO doesn't has the potential ability to appropriately handle the relationship between exploitation and exploration and hence the maximum possibility to converge at local minimum. The aforementioned drawbacks have been overcome through various modifications such as quantum-behaved PSO, CPSO, etc. Due to various modifications, there are now numerous variants of PSO and hence it has become impossible to allow a new proposed PSO to go through all the test functions so that comparison can be made with the other proposed PSO variants. Hence a common platform for PSO wherein used can submit their versions of PSO and then a comparison can be made so as to reveal the performance.

The application potential of PSO was, however, impaired owing to the lack in depth theoretical aspects. Therefore a more comprehensive theoretical study of both the run-time and convergence properties associated with the PSO algorithm is required to be carried out.

A number of applications associated with mechanical engineering domain have been depicted toward the end of the chapter.

PART II

NATURE-BASED TECHNIQUES

4 Ant Colony Optimization

4.1 INTRODUCTION

Ant colony optimization (ACO) algorithm was proposed by Dorigo (1992). ACO was inspired by ants that finds the shortest path from their nest to the source of food. They do this by laying pheromone on the ground during their return trip to the nest. The other ants then follow the pheromone on the ground to reach the food source from the nest. If there are two trails of pheromone, then the ants will follow the one with heavier pheromone layer with higher probability. For instance, suppose round trip is made by some ants to the food source within a minute along one of the pheromone trail, say A. The ants, on another pheromone trail B, take around 2 min. Initially both the trails of pheromone have equal heavy pheromone layers. Let there be 100 ants that depart for the food source. Out of these 100, 50 choose trail A. After 1 min, these 50 ants will return having augmented trail A. However, trail B is still to be augmented. Hence, this time more than 50 ants will choose trail A again. Now after 2 min, more than 75 ants would have augmented trail A and only 50 trail B. This iterative process will lead all the ants to choose trail A i.e., the shortest route. The same feedback mechanism applies to the subpaths.

The built-in-optimization capability of ants has been revealed through the double bridge experiments whereby an ant finds the shortest possible path between the two points in the environment. This is done through the employability of certain probabilistic rules that are based on the available local information (Dorgio and Stutzle, 2004).

ACO is one of the prominently used metaheuristics wherein good solution to an optimization problem is found through the cooperation from the colonies of artificial ants. As such dynamic and static problems are solved using ACO algorithm both in discrete as well as continuous domains. Simple agents are augmented with the computational resources that communicate with other ants indirectly through the pheromone trail. The transitions between different iterations are made through probability distribution concept. The potential ability of ant algorithms to solve discrete combinatorial optimization problems has already been evidenced, and recently a number of reports on the successful application of ACO to continuous search space have also been reported.

The artificial agents are stochastic constructive agents that move on the construction path and therefore builds solution. The constructive aspect of ACO distinguishes it from other optimization algorithms. Each ant incepts from an initial solution state and then build its solution or a component of solution. Each ant also collects information in addition to building its own solution. The information collected is associated with the problem characteristics as well as its own performance. The information collected aids in modification of the problem representation in accordance with the perception of the other ants in the search space. A finite sequence of neighbor states is employed by each ant to build the solution. A stochastic local search policy aids in making the selection to the moves. This is done by the information possessed by the individual ant, the available pheromone trail and through the employability of predefined problem-specific information. A long-term memory of the entire search process is encoded by the pheromone trail which is updated by the ant themselves. Heuristic information is also provided with due consideration to priori information about the problem instance and by a source different from the ants in the search space.

ACO has three main phases: construction of solution, updating the pheromone trail and daemon action. Artificial ants move through the adjacent states of the problem during the construction space in accordance with the transition rule and then builds on the solution iteratively. The pheromone trail are updated in the second phase of ACO by both the pheromone trail evaporation and reinforcement. The third phase i.e., the daemon action is optional and is about application of additional updates from a global perspective (Jalali et al., 2007a). The characteristics of the problem define on when and how the ants should release pheromone on the environment. The design of the implementation also plays a critical role in the same. The solution can be built by releasing the pheromone in a step-by-step procedure and the release can be postponed until the solution is built completely.

The search is directed to the most interesting regions in the search space, a stochastic functional composition of the pheromone trail available locally. An immature convergence, wherein all the ants may drift toward the same region of search space, may be avoided through the presence of stochastic component and pheromone evaporation mechanism. The level of stochasticity in the policy can control and therefore provide a balance between the exploration of new search points and the exploitation of the accumulated knowledge. The ant is deleted from the system, once the complete solution has been built and the pheromone has been updated.

A number of variations to ACO have also been developed and will be discussed in the present chapter. The basic components and the goal of ACO have been discussed next. This is followed with a discussion on various variants of ACO.

4.2 COMPONENTS AND GOALS OF ACO

This section of the chapter delineates the different components of ACO algorithm. The role of ACO method has also been discussed.

Pheromone: Pheromone is a chemical factor that an ant excretes and therefore benefits the other ants in the search space. These pheromones therefore aid in finding the optimal solution in several optimization problems of combinatorial and uncertain nature. The pheromone trails have volatile nature and get evaporated after certain amount of time. The artificial ants also have evaporation factor that helps in defining priority for non-optimal paths.

Initial pheromone value: This signifies the value of the pheromone laid by ants. Initial value of the pheromone trail can be set either 0 or 1. If the value assigned is 1, then the rate of evaporation will be set such that the pheromone gets first evaporated and then the pheromone is laid over the evaporated path. If value of 0 is selected, then first the pheromone will be laid and then the evaporation factor will reduce the quality of pheromone deposited on the path.

Stigmergy: Indirect communication between the ants via the deposited pheromone is referred to as Stigmergy. The successor ant takes note of the pheromone concentration deposited by the previous ant and then decide on as to which path is to be followed. Therefore this method of indirect communication is known as Stigmergy.

Rate of pheromone evaporation: The decay of pheromone in a unit time interval is known as pheromone evaporation rate. Artificial ants require this evaporation so as the non-optimal path can be avoided to be trace by the successor ants. This means speeding up the rate of convergence. The evaporation rate is represented by Δ.

Coefficient of pheromone decay: Decay of pheromone is constant and hence the coefficient of pheromone decay.

ACO has the objective to solve problems that are hard in nature. Number of problems can be handled by ACO, even the NP hard problems that can be solved only in polynomial time. Such kind of solution require only the best solution and not the exact solution. ACO can be explored as such for solving such problems.

The artificial ants are employed in order to find the optimal path between the source nodes to the destination node. The behavior of real ants are emulated by the artificial ants. The path build at first to the destination is not worse and is optimized subsequently through the pheromone trail. The same behavior is emulated by the computational ants and the optimality for any problem is deciphered. Pheromone laid on the ground also has certain evaporation rate so that the non-optimal path gets omitted through the evaporation phenomenon. Real ants used to move to the food source by crossing some of the intermediary adjacent nodes, this aids them to reach the food source at a faster rate.

The path chosen by real ants can be accomplished in two ways: The path traversed by the previous ants can be traveled by the present ants with the aid of pheromone laid. The other way is that the ants by some means of heuristics choose to follow a path in random manner from their source to the destination. Similar procedure is adopted by the real ants that move from a node i to node j i.e., it can follow the path created in the previous iteration or it can choose a random path in accordance with the program. The artificial ants also have other qualities that are not possessed by the real ants. The capability to portend the future path is only possessed by the artificial ants. The real ants always follow a path on the basis of laid pheromone trails or in some random manner, but the artificial ants can always tread a path that can ultimately result in reaching an optimal solution. As such the optimality of solution produced is better in comparison to the real ants.

Artificial ants also possessed internal memory that aids them to store their past actions that have been made by them in the previous iterations as well as the present ongoing iterations. The memory is useful in updating of the pheromone trail. The pheromone update will take place once the ants construct the complete tour. The internal state comes in handy while updating the path from the destination to the source.

4.3 TRADITIONAL APPROACHES OF ACO

The ACO was introduced in the year 1991 and since its inception a number of algorithms have come into existence. The following section of the chapter illuminates the reader with a number of such algorithms.

4.3.1 ANT SYSTEM

The ant system was proposed by Dorgio (Mohan and Baskaran, 2012; Colorni et al., 1992; Vijayaragavan et al., 2013) and was applied to solve the traveling salesman problem. The problem considers to move the ant from one node in the graph to the other. The algorithm was made to run t number of times. An m number of ants build a tour executing n number of steps wherein the state transition rule is applied. Initial inception of the ant system carried three versions: ant cycle, ant quantity and ant density. There is a notable difference between the first version and the last two versions. In the first version the update using the trail of pheromone is done at the edges only when the ant complete tour has been constructed. While in the latter two versions, the ants deposit pheromone in all the paths traveled between node i and node j. In the first version the pheromone will be laid at the edges only when the quality is found to be better than the previous tour. The first version provides better results in comparison to the other two. As such the first version was considered and the last two were abolished.

Tour construction and pheromone update are the two main steps for any ant system.

In the *tour construction* phase, each ant will be put on to some random chosen city. A state transition rule will be given to each ant k to complete the tour. Equation (4.1) depicts the probability of the ant k to move from city i to j:

$$p_{ij}^k(t) = \frac{\left[\tau_{ij}(t)^\alpha \cdot [\eta_{ij}]\right]^\beta}{\sum_{l \in N_i^k}\left[\tau_{ij}(t)^\alpha \cdot [\eta_{ij}]\right]^\beta} \quad \text{if } j, \eta \in N_i^k \tag{4.1}$$

where, $\eta_{ij} = 1/d_{ij}$ is the available heuristics, the kneeboard node for ant k is denoted by N_i^k, for $\alpha = 0$ the neighboring cities will be selected and the pheromone amplification will work if $\beta = 0$.

Pheromone update comes into play once the ants have finished to construct the path. The pheromone update will be done after decreasing the strength on entire arc by constant factor and then the pheromone deposition will take place in accordance with the Equation (4.2):

$$\tau_{ij}(t+1) = (1-\rho) \cdot \tau_{ij}(t) + \sum_{k=1}^{m} \Delta\tau_{ij}^k(t) \tag{4.2}$$

where, ρ is the evaporation of pheromone trail and ranges 0 and 1, $\Delta \tau_{ij}^k(t)$ represents the pheromone deposition amount on the arc it traveled. This is defined using Equation (4.3):

$$\Delta\tau_{ij}^k(t) = \begin{cases} \dfrac{1}{L^k(t)} & \text{if arc}(i,j)\text{ is used by ant } k \\ 0 & \text{otherwise} \end{cases} \tag{4.3}$$

In the above equation, the length of tour for ant k is represented by $L^k(t)$.

4.3.2 MAX-MIN ANT SYSTEM

The solution in Max-Min Ant System (MMAS) (Dorigo and Gambardella, 1997; Gambardella and Dorigo, 1996) is constructed in the same way as in the ant system. There are three qualifying steps in MMAS. The first step in the MMAS technique is to find the best solution found during the iteration, and after each iteration the pheromone update is allowed to be done by only one ant. There are two possibilities of ants that can aid in updating pheromone trail: iteration best ant and the global best ant. The iteration best ant is the one that finds the best solution in the current iteration, whereas the ant that finds the solution from the beginning of trial is referred to as global best ant.

In order to avoid stagnation of the search space, the trail of pheromone is limited in the interval $[\tau_{min}, \tau_{max}]$. Thirdly, the pheromone trail will be set initially to τ_{max} which aids in maximizing the search space for exploration.

The pheromone trail will be updated in accordance with Equation (4.4), once all the ants construct the solution:

$$\tau_{ij}(t+1) = (1-\rho)\cdot\tau_{ij}(t) + \rho\cdot\Delta\tau_{ij}^{best} \tag{4.4}$$

The ant that is required to update the pheromone may be either iteration or global best ant. The arc will participate in the global best solution, if the arc is traversed often by the ant. This is because the most traversed arc will receive the most pheromone. The use of global best solution or the iteration best solution has an important reason i.e., when the global best solution is employed to update the path, there may be possibility that the search may be focused around this global best solution. This will limit the exploration of much better path and as such the algorithm may get trapped into poor quality solutions. This can, however, be avoided if the iteration solution is adopted to update the pheromone trail. This is because the solution may differ iteratively and as such development of many arcs may take place and hence can be explored.

Lower and upper limits for pheromone trails are the characteristic features for the MMAS algorithm. These trail limit will aid in avoiding any path to be traversed by the ants that fall beyond the trail intervals. The elimination is required as that particular arc may lead to good quality of solution.

4.3.3 QUANTUM ANT COLONY OPTIMIZATION

The main motivation behind quantum ant colony optimization (QACO) (Vijayaraghavan et al., 2012) is the introduction of quantum rotation gate and Q-bit representations into ACO. The major objective to build a discrete binary ACO algorithm and the implementation of hyper-cube framework.

Working of QACO comprises of following steps: the initial pheromone can be represented using Equation (4.5), given the fact that initial parameters have been defined as such $\alpha = \beta = 1/2^{1/2}$

$$\tau = \begin{bmatrix} \tau_{1\alpha} & \tau_{2\alpha} & \cdots & \tau_{m\alpha} \\ \tau_{1\beta} & \tau_{1\beta} & \cdots & \tau_{m\beta} \end{bmatrix} \tag{4.5}$$

A random parameter, p, is generated and is compared with the probability parameter, p_e.

If $p < p_e$, Equation (4.6) can be used to find the solution of ith and jth bit:

$$\text{solution}_{i,j} = \begin{cases} 0 & \text{if } \tau_{j \cdot \beta} \leq \tau_{j \cdot \alpha} \\ 1 & \text{if } \tau_{j \cdot \beta} > \tau_{j \cdot \alpha} \end{cases} \tag{4.6}$$

If $p > p_e$, Equation (4.7) can be used to find the solution of ith and jth bit. This is done using the threshold function.

$$\eta_c(x) = \begin{cases} 0 & c < \eta_0 \\ 1 & c \geq \eta_0 \end{cases} \tag{4.7}$$

In the next step, calculation of the best function is accomplished and then checked for the termination condition. If the termination condition is not satisfied, then the pheromone density will be updated in accordance with the following:

$$\begin{bmatrix} \tau'_{1\alpha} \\ \tau'_{1\beta} \end{bmatrix} = R(\theta_i) \begin{bmatrix} \tau'_{1\alpha} \\ \tau'_{1\beta} \end{bmatrix} = \begin{bmatrix} \cos\cos(\theta_i) & -\sin\sin(\theta_i) \\ \sin\sin(\theta_i) & \cos\cos(\theta_i) \end{bmatrix} \cdot \begin{bmatrix} \tau_{1\alpha} \\ \tau_{1\beta} \end{bmatrix} \tag{4.8}$$

where, θ is the rotation angle.

4.3.4 COOPERATIVE GENETIC ANT SYSTEM

This algorithm has some similarities with respect to hybrid algorithms such as the algorithm combining genetic algorithm and the ant system. The cooperative genetic ant system (CGAS) algorithm (Bai et al., 2013) executes both the genetic algorithm as well as the ant system cooperatively and simultaneously. Therefore, with the CGAS algorithm it is better to achieve good solution for different problems such as the traveling salesman problem. The CGAS algorithm was first applied for the traveling salesman problem.

Ant system is employed for the initial construction of the tour. Genetic algorithm is then initialized through the aid of constructed tours. The ant system is then used for the construction of new solutions and genetic algorithm is employed for generation of new genes. The global best solution will be chosen, among the two set of best solutions. The pheromone update then will be done. The algorithm will end, once the termination condition has been satisfied.

The CGAS algorithm selects the best city considering the following equation:

$$j = \{ \min c(i) \quad \text{if } j \in s_k \tag{4.9}$$

where, i the current city, j is the next city to be visited and $c(i)$ is the sorted list of cities to be visited.

4.3.5 CUNNING ANT SYSTEM

The cunning ant system algorithm is different in comparison to the traditional ACO technique. It is manner of construction of solution that differentiates the cunning ant system from the traditional ACO technique. The solution is constructed by borrowing a part of solution from the already existing solutions. The remaining part of the solution is built by following a usual traditional ACO algorithm.

An ant is known as a donor ant if it provides solution to the ant that is ready to borrow the solution. The ant borrowing the solution is known as cunning ant. Therefore at any iteration t, a borrower ant, known as cunning ant (c-ant) borrows a part of solution from the donor ant (d-ant). In the subsequent steps, the c-ant$_{k,t+1}$ is compared with d-ant$_{k,t}$ and the best solution among the two is selected and is left to be continued for the next iteration. The update of pheromone will then take place in accordance with the following equation:

$$\tau_{ij}(t+1) = \rho \cdot \tau_{ij}(t) + \sum_{k=1}^{m} \Delta * \tau_{ij}^{k}(t) \tag{4.10}$$

$$\Delta * \tau_{ij}^{k}(t) = \begin{cases} \dfrac{1}{C_{k,t}^{*}}, & \text{if } (i,j) \in \text{ant}_{k,t}^{*} \\ 0, & \text{otherwise} \end{cases}$$

where, fitness value is represented by $C_{k,t}^{*}$ for any ant signified by ant$_{k,t}^{*}$.

The maximum value of pheromone trail in case of the cunning ant system can then be derived using the following equation:

$$\tau_{\max}(t) = \frac{1}{1-\rho} \times \sum_{k=1}^{m} \frac{1}{C_{k,t}^{*}} \tag{4.11}$$

and

$$\tau_{\min}(t) = \frac{1}{1-\rho} \times \frac{1}{C_{t}^{\text{best-so-far}}} \tag{4.12}$$

where, the fitness value of best-so-far solution at any iteration t is denoted by $C_{t}^{\text{best-so-far}}$.

4.3.6 MODEL INDUCED MAX-MIN ANT SYSTEM

The model induced max-min ant system is one of the hybrid optimization algorithm that was developed initially to solve asymmetric traveling salesman problem. The complex analytical results obtained from the asymmetric traveling salesman problem is combined with the max-min ACO algorithm. The model induced max-min ant system contributes in the following ways to obtain the best possible solution for the problem under consideration. Statistic biased weighting factors are replaced with the dynamic ones and adjusted transition probabilities are built. The replaced dynamic weighting factor is dependent closely on the partial solution build and developed by the ants. The main idea behind this scheme is that it favors the choices that has lesser residual cost instead of small actual cost. Non-optimal arcs are identified as a result of by-product and at each step of the construction of the path. This is accomplished employing the dual information. Associated assignment problem is the main source of such dual information and ultimately aids in discarding the arcs from the future consideration.

Secondly, the contribution lies in the determination of terminal conditions analytically. This is based on the state of pheromone matrix structure. The results obtained comes with the necessary conditions to obtain optimal solution.

Model induced max-min ant system algorithm employs the assignment problem to arrive at residual cost and the solution is repaired using the PATCH algorithm. The solution repaired is given by the assignment problem to deduce the lower bound. The lower bound residual cost matrix is then established with the aid of lower bound. The first candidate solution for the model induced max-min ant system is returned with the employment of PATCH algorithm. The termination condition t_0 will be then set and then the minimum pheromone value is set in accordance with the following equation:

$$\tau_{min} = \frac{1}{\left(\frac{1}{2}\right)f(s_1) + \left(\frac{1}{2}\right)Z_{AP}^*} \tag{4.13}$$

where, residual cost matrix is represented by Z_{AP}^*. After all the derived things, the value of pheromone will be calculated and set dynamically. The tour will then be constructed and then the local search will be employed by adopting the 2-OPT heuristics. Best solutions as well as the gap will also be updated.

4.3.7 ANT COLONY SYSTEM

Dorgio in 1997 developed the ant colony system (Dorigo, 1992; Stützle and Dorigo, 1999). There are three factors that distinguish ant colony system from the ant system. The first distinction lies in that the ant colony system employs more aggressive action in comparison to the ant system. Secondly, the pheromone is updated only to the arcs and associates itself to the global best solution. Thirdly, some pheromone gets deduced whenever the ant moves from a city i to city j. The deduction is constant in nature. The detailed discussion on the alterations follows next.

A pseudo random proportional action choice rule is adopted by the ant colony system to choose the next city as opposed to the transition rule employed in case of the ant system. An ant k moves from city i to city j with a probability of q_o. That means the move from node i to node j will be guided by the heuristics or by updating of the pheromone trail. The ant performs exploration of arcs with the probability of $(1-q_o)$.

Now a discussion on the alterations in update of global pheromone. In the ant colony system, pheromone gets updated after each iteration and hence the global best solution will also get updated. The update of the pheromone trail will be done by the global best ant. The identification of the global best solution is done on the basis of the solution returned after each iteration. The update formula is given by the following equation:

$$\tau_{ij}(t+1) = (1-\rho) \cdot \tau_{ij}(t) + \rho \cdot \Delta\tau_{ij}^{gb}(t) \tag{4.14}$$

The local pheromone update will be done after each ant travels the arc. The update will change the quantity of the pheromone on the arc. The quantity of pheromone will aid the next ant to choose the pheromone trail accordingly. This ultimately leads to exploration of number of arcs. The local update of the pheromone trail is done in accordance with the following equation:

$$\tau_{ij} = (1-\xi) \cdot \tau_{ij} + \xi \cdot \tau_0 \tag{4.15}$$

where, ξ and τ_0 are the two important parameters, wherein ξ ranges between 0 and 1.

4.4   ENGINEERING APPLICATIONS OF ANT COLONY OPTIMIZATION ALGORITHM

A number of reports have been published on the optimization of different non-traditional machining processes. Teimouri and Baseri (2014) carried out multiple objective optimization of the electrical discharge machining (EDM) process by considering SPK (X210Cr12) cold work steel as the material for the tool. ACO was employed to optimize the objective function as a function of surface roughness and material removal rate. The process parameters considered were: rotational speed, magnetic field intensity, pulse-on time and its product with product current. Wire EDM (WEDM) process was also optimized using the ACO technique (Mukherjee et al., 2012). Both single as well as multiple objective optimization were considered and such as material removal rate, surface roughness, wear ratio and kerf width were considered to be the objective functions. Peak current, pulse frequency, duration of the pulse, flow rate of the dielectric, wire tension, wire speed and water pressure were considered as the process parameters.

4.5   CONCLUSION

Present chapter has provided a brief overview of the ACO technique. The different variants of the aforementioned optimization technique have been considered. A number of applications in the domain of non-traditional machining process have been discussed toward the end of the chapter.

5 Bees Algorithm

5.1 INTRODUCTION

There have been many metaheuristic algorithms that have been developed on the basis of their inspiration from collective intelligent behaviors of swarms of insects and animals. Such intelligent swarms have the potential ability to achieve fascinating complex behaviors irrespective of the behaviors of their individuals (Garnier et al., 2007). Self-organization possessed by such animals and insects is the basis behind the fascinating complex behaviors endorsed by them (Bonabeau et al., 1999). Self-organization is a characteristic feature wherein the organization is exhibited without the organizer. There is no need for any internal or external guidance for such characteristic behavior. The social beings update their activities by themselves through the aid of decentralized control mechanism. The decentralized mechanism involves information from local and limited sources. The fascinating intelligent collective behavior as well as the capabilities shown by the social animals has attracted various researchers to solve the daily life optimization problems. The models can be employed for the development of artificial versions. This is done either by tuning in the parameters of the model through the values outside the biological range or through the assumption of additional non-biological characteristics. As a result of such modifications or assumptions, the biological swarm intelligence has been transformed to artificial systems. This gave rise to swarm intelligence under the domain of artificial intelligence. As such many algorithms have been developed: ant colony optimization (ACO), particle swarm optimization (PSO), firefly algorithm and bacterial foraging optimization (BFO).

Honeybees are among the social beings that depicts fascinating features and behaviors. When the honeybees are in swarms, they exhibit a number of surprising intelligent behaviors. Researchers also have get attracted toward the collective intelligent behavior of swarm of honeybees and therefore have developed number of intelligent search algorithms. Examples of such algorithms include: artificial bee colony (ACO) (Karaboga and Basturk, 2007b), bee colony optimization (BCO) algorithms (Davidovic et al., 2014; Teodorovic et al., 2015) and honeybee mating optimization (HMBO) (Abbass, 2001). A number of comprehensive surveys have been conducted by number of researchers on the swarm intelligence of bees (Karaboga and Akay, 2009a; Diwold et al., 2011).

Pham et al. (2006a) proposed the Bees Algorithm. The prosed algorithm has been inspired from the foraging behavior of the honeybees for their food. The Bees Algorithm performs a kind of neighborhood or local search and is combined with exploratory global search. A uniform mode search is adopted by both the types of search modes. In case of global search, there is a uniform distribution of the scout bees across the search space. These scout bees search for the potential solutions in the search space. Follower bees, on the other hand, are recruited in case of the local or neighborhood search. These are employed for the patches identified by the scout bees and these bees exploits these patches. Bees Algorithm has been employed for a number of optimization problems such as job scheduling problems (Pham et al., 2007a; Yuce et al., 2015a), timetabling problems (Abdullah and Alzaqebah, 2013; Lara et al., 2008) and training of neural network (Pham and Darwish, 2010).

Bees Algorithm has attracted a lot of attention from researchers for its simplicity and closeness to the actual behavior in the biological system. There are four different components associated with the Bees Algorithm: tuning of the parameter, initialization, global search and local search. There have been numerous studies on enhancing the performance through improvements in the critical components. Therefore a number of variants of Bees Algorithm have emerged since the inception of the Bees Algorithm.

The present chapter provides an overview of basic main implementation of the Bees Algorithm. Furthermore, the most important improvements made in the Bees Algorithm have been discussed. The applications of Bees Algorithm are presented toward the end of the chapter.

5.2 BASIC VERSION OF BEES ALGORITHM

Bees Algorithm has been inspired by the foraging behavior of the honeybees in nature. A bee colony, during the harvesting season, employs part of its population to scout the surrounding areas around the hives (Tereshko and Loengarov, 2005). Scout bees randomly searches for the food in the surrounding areas. Flower patches with abundant nectar are searched in particular by the scout bees. As such the extraction is easier and the patches are rich in sugar content.

Scout bees deposit the collected nectar during the search process. The scout bees that have been able to identify the high-quality food source, signals the positions of their identification to the resting mates. This is done through a ritual known as the *waggle dance* (Von Frisch, 2014). The ritual of waggle dance is performed in a particular area of the hive which is known as the *dance floor*. Through this ritual, mainly three pieces of information are communicated: the direction of the located flower patch, the distance of the identified flower patch from the hive and its quality rating. The bees then go back to the flower patch after the ritual of waggle dance is finished. The other mates from the hive follow the bees on their way to the flower patch. The quality rating determines the number of bees to be recruited. Largest number of foragers are attracted by the flower patches containing richer and easily available nectar or the pollen sources. The recruiter bees will then inspire other bees in the hive through the waggle dance to follow them to the flower patch. As such the most profitable food source attracts the maximum number of foragers (Tereshko and Lee, 2002). Therefore the efficiency of the food collection process is optimized by the bees i.e., the amount of collected food to the harvesting cost.

Each point in the search space represents a food source and hence a potential solution. Random sampling of the search space i.e., generation of random solutions, is done by the scout bees. The quality of the searched space is then reported via the fitness value. In a way, the quality of the generated solution is evaluated through the process of quality evaluation. The ranking of the sampled solutions is performed and the other bees in the hive are recruited to search the space around the highest ranked landscape i.e., other solutions are explored within the neighborhood of the highest ranked location. The neighborhood of the solution is referred to as *flower patch*. Most promising of all solutions is explored by the Bees Algorithm and it selectively explores their neighborhood in search of global minimum of the objective function. The details of the Bees Algorithm is now presented.

Representative scheme: Let the space of problem solution be represented as $U = \{x \epsilon R^n; \max_i < x_i < \min_i i = 1,\ldots,n.$ the fitness function $f(x)$: $U \rightarrow R$ and each candidate solution is expressed as n-dimensional vector of decision variables $x = \{1,\ldots x_n\}$.

Initialization: The population of scout bees is represented by ns and is fixed. However, the population of scout bees is randomly dispersed across the search space with uniform probability. Fitness function is employed to evaluate the quality of the visited site. The main loop of the algorithm then incepts and comprises of four phases. On meeting the termination criteria, the algorithm stops.

Waggle dance: The fitness information collected by the scout bees is used to rank the sites visited by the scout bees. The local exploration then proceeds with the nb number of highest ranked locations. The other bees perform the local exploration that are directed by the scout bees to the neighborhood of the selected sites. The number of foragers for each selected site is determined deterministically as follows: waggle dance is performed by the scout that returns from the best nb number of sites. Therefore, the nest mates are employed for carrying out the local exploration. There are certain number of elite sites i.e., ne among the best nb sites and hence nre number of foragers are employed to search in the neighborhood of the elite sites.

As such more and more bees are employed to carry out the search in the solution space in the vicinity of the ne points carrying highest fitness value. The local search operation is therefore more thorough in the vicinity of the elite sites. The fitness-based differential recruitment is therefore one of the key operations for the Bees Algorithm.

Local search: The recruited bees are placed randomly at each of these nb selected sites. These bees are placed with a uniform probability on the neighborhood of the sites with highest fitness values identified by the scout bees. An n-dimensional hyper box of sides $a_1,\ldots,a_n$ defines the neighborhood with the scout bees at its center. The fitness value is again evaluated for each of the flower patches visited by the recruited bees. If the recruited bee lands itself in a position where the fitness value evaluated is higher than that of the scout bees, then the recruited bee assumes the role of the scout bee. At the end of the process, only the bees with the better fitness value is retained. The bees then become the dancer back at the hive. However, in the biological system, the feedback differs since the waggle dance will be done by all the bees that are involved in the foraging process.

Global search: ns-nb bees are placed across the fitness landscape in the global search phase. These are placed randomly across the identified fitness landscape. That scouts for the new patches of flower. The exploration effort of the Bees Algorithm is revealed through the random scouting process.

Population update: A new population of the bee colony is established at the end of each iteration. The new population comprises of two groups: there are nb scout bees associated with the first group and forms the center of the flower patch. This group represents the results obtained as a result of local exploitive search. On the other hand, *ns-nb* bees form the second group that associate itself with the randomly generated solution and hence is representative of the global explorative search.

Stopping criterion: The domain of the problem under consideration defines the stopping criterion and as such it can be the location of a solution of fitness value above a certain threshold or it may be the completion of a predefined number of evolution cycle.

5.3 IMPROVEMENTS ON BEES ALGORITHM

Bees Algorithm has been applied to a wide range of problems such as solving time-tabling problems (Nguyen et al., 2012; Abdullah and Alzaqebah, 2013; Lara et al., 2008), image threshold (Hussein et al., 2016; Shatnawi, 2013; Azarbad et al., 2011), container loading problems (Dereli and Das, 2011), optimization of printed circuit board assembly (Ang et al., 2010; Pham et al., 2007b), protein conformation search (Bahamish et al., 2008), inverse kinematics of a robot manipulator (Pham et al., 2008a), finite element model updating (Moradi et al., 2010), recognition of handwritten digits (Nebti and Boukerram, 2010), classification of wood defect (Packianather and Kapoor, 2015), optimization of supply chain (Mastrocinque et al., 2013; Yuce et al., 2014, 2015b), planning of minimum time motion for robot arm (Ang et al., 2009) and other optimization problems.

Bees Algorithm has attracted significant level of interests from the research community because of its simplicity and closeness to the biological behavior of the bees. There are four parts into which the Bees Algorithm can be divided: tuning of parameters, initialization, local search and the global search. There have been umpteen studies to improve the Bees Algorithm and hence enhance its performance. The following sub-sections illuminates the readers with different improvements.

5.3.1 IMPROVEMENTS ASSOCIATED WITH SETTING AND TUNING OF PARAMETERS

Some of the studies have been concentrated to tune the parameters and their setting. The associated studies were made either by reduction in the number of tunable parameters (Pham and Darwish, 2008) or by development of the tuning methods and therefore tune the parameters of the basic Bees Algorithm (Imanguliyev, 2013; Otri, 2011) or by tuning the extra parameters associated with the modified variants of the algorithm (Pham et al., 2012). The tuning process is one of the important aspects of any metaheuristic as it aids in enhancing the performance of the optimization technique. However, there is no exact method for tuning the parameters of a metaheuristic algorithm and as such that can aid in determination of the parameter values that are most suitable for all the classes of problem. Manual tuning by hit and trial is the most common tuning method. However, manual tuning method is time expensive.

Otri (2011) has proposed a tuning approach known as the Meta Bees Algorithm that helps in finding the suitable values of the parameters while solving the problem by employing the Bees Algorithm. The Meta Bees Algorithm is composed of two different parts: wrapper Bees Algorithm and the wrapped Bees Algorithm. Pilot test aids in determination of the parameter value associated with wrapper Bees Algorithm. A set of parameter value is represented by each of the bee of wrapper

Bees Algorithm which are used by wrapped Bees Algorithm to solve the optimization problem. The parameters are initialized within a chosen range. At each iteration, wrapped Bees Algorithm is executed and is continued until the stopping criteria is met. Optimization problem is used by the wrapped Bees Algorithm as a fitness function. On the other hand, the number of function evaluations are employed by the wrapper Bees Algorithm as a fitness function and therefore measure the quality of different parameter settings. Hence Meta Bees Algorithm is employed to determine suitable values of the parameters required to solve the optimization problem.

Imanguliyev (2013) proposed a self-guiding tuning method that requires lesser manual interaction in tuning the parameters of the basic Bees Algorithm. In the proposed tuning method, predefined values are employed as default values of the parameters. Then on the basis of previous experiences, the maximum values of the independent parameters are decided. Two steps are used in tuning the parameters one at a time: fine and rough tuning. The range limited by the maximum parameter value is divided into number of groups for the rough tuning process. The problem under the study is optimized by selecting a random number from each group. The group of numbers that provides the most accurate results with the smallest number of evaluations will be selected to have performed the fine-tuning. All the integers in the group are involved in fine-tuning process. The number that provides the best result with smallest number of iterations will be chosen finally as the value of the tuned parameter. Next parameter is tuned, once a given parameter has been fine-tuned.

A selection mechanism based on the fuzzy greedy system was introduced in the selection mechanism by Pham and Darwish (2008) in the local search phase of the basic Bees Algorithm. Reducing the number parameters required to run a basic Bees Algorithm was the objective behind the aforementioned introduction. This was achieved by automating the selection and the recruitment process. The number of sites selected as well as the number of recruited bees are selected automatically through the aid of fuzzy rules. The usage of fuzzy rules aids in selecting the patches on the basis of certain factors and at the same time provides information on the number of bees to be recruited for the selected patches. The factors are based on the criterion behavior of the system and represents the condition in the "IF" part of the fuzzy rules. The two factors: fitness and rank are used if the system has single-criterion behavior. On the other hand, in case of a multi-criterion system wherein the fitness of the visited patches is represented, the involved criterion forms the factor in the fuzzy rule.

5.3.2 Improvements Considered on the Local and Global Search Phase

There have been studies on improving the performance of the Bees Algorithm through the improvements in the local search part (Yuce et al., 2013; Ahmad et al., 2012; Packianather et al., 2009; Pham and Castellani, 2009) or through the improvements in both the global as well as the local search parts (Imanguliyev, 2013; Shatnawi et al., 2013; Pham et al., 2012; Ghanbarzadeh, 2007).

Two procedures were proposed by Ghanbarzadeh (2007) to make improvements in the local search stage of the basic Bees Algorithm. Site abandonment as well as neighborhood shrinking were the two adopted procedures. The neighborhood size, *ngh*, is initialized to a large value in case of the neighborhood shrinking procedure.

The size of the patch is maintained until and unless the local search process yields solutions with higher fitness value. The size of the neighborhood is shrunk when the local search process fails to produce the desired candidate solution. The decreased neighborhood size makes the local search process more exploitive and as a result increases the density of the sampled solution around the fitness landscape. The flower patch site, if doesn't produces any significant improvement through the application of shrinking procedure for subsequent iterations, is abandoned. The procedure is known as the abandonment procedure and the abandoned site is replaced with a newer site, even though the new site has lower fitness. The replaced site is stored temporarily before the abandonment procedure. The replaced site as such can be retrieved unless a better solution is identified. If the patch produces no better solution even after subsequent iterations in the local search phase, then the patch is centered on the local minima. In such cases the abandonment of the patch represents an escape from the local optimum.

In another proposed improvement by Ghanbarzadeh (2007), a modification to the Bees Algorithm was achieved by improvising the global search part. This was done by mating the selected bees exploiting the best patches with the unselected bees that explore the space globally. The interpolation as well as extrapolation mating methods are employed by the unselected bees to improve their positions. By doing so, the unselected bees don't reposition themselves randomly. Each unselected bee mates with the selected bee and then extrapolate or interpolate along with the selected bee. As such the unselected bee repositions itself in accordance with the information received from the selected bee. The repositioning is done in between the current position and the position of the selected bee in case the unselected bee repositions by interpolation. While in the case of extrapolation repositioning, expansion of the line segment joining the selected and the unselected bee is done in both the directions. The expansion is done by half of the length of the line segment. The repositioning is done by the unselected bee at either of the expanded segments.

Yuce et al. (2013) **improvised** the local search part of the Bees Algorithm through an adaptive neighborhood size change and the strategy of site abandonment. The conceptual idea of site abandonment as well as the neighborhood shrinking proposed by Ghanbarzadeh (2007) was applied in a distinct manner. As such the shrinking strategy is applied after no obvious improvements are visible in the local search even after subsequent iterations. An enhancement strategy is then applied after no obvious improvements from the application of the shrinking procedure is achieved after successive iterations. The enhancement strategy is applied for the iterations wherein no progress is observable. The site is unproductive and is finally it is abandoned. The position at this instance is known as the final solution and in such cases no better solution can be achieved any further.

An improvement in the basic Bees Algorithm was proposed through the recruitment mechanism by Packianather et al. (2009). The recruitment mechanism is based on the pheromone communication system that is possessed by the honeybees in nature. Automation of the recruitment process in the basic Bees Algorithm and therefore make it a dynamic process. The allocation of the bees is not solely dependent on single parameter as is the case with the basic Bees Algorithm. The level of the pheromone decides the number of bees to be recruited. Each bee deposits pheromone on one of the selected sites and the amount of pheromone deposited depends on the fitness of the site under consideration. Also the pheromone deposited earlier and the number of bees searching the site also dictates the number of bees that are required to be recruited. However, with the time, the pheromone deposited gets evaporated and as such the site loses its attractiveness as it has been exploited well and the nectar present in the flower might have depleted. Hence with time, the number of bees to be recruited decreases over time. The dynamic nature of the Bees Algorithm is thereby revealed because of the changing scenario of deposition, and evaporation of the pheromone as well as the number of recruited bees.

Pham and Darwish (2008) proposed a fuzzy greedy selection mechanism and improved the basic Bees Algorithm. The authors combined their mechanism with Kalman filter (Zhang and Wunsch, 2003). The combined mechanism is a recursive estimator that aids in the prediction of the optimal parameters for the linear as well as the nonlinear systems. In the basic Bees Algorithm, it updates the position of the follower bees in the local search phase of the algorithm. The Kalman filter also aims to enable the migration of the Bees Algorithm toward a good solution. Kalman filter gain and estimation error aid in controlling the extent of the local search area. Therefore instead of random search in the neighborhood, the algorithm adopted the fuzzy greedy selection mechanism, Kalman filter and the site abandonment procedure.

Levy flights was introduced by Shatnawi (2013) for improvising the initialization, local and global search parts of the memory-based Bees Algorithm. The memory-based Bees Algorithm was proposed by Shatnawi (2013). There are certain issues that need to be done for the improvising the initialization part of the Bees Algorithm. Also, the Levy flights were modeled by Shatnawi (2013) for improvising the global search but it was not modeled for the distributed patch in the nature. For the local search part, the Levy flights were modeled for initialization, and the global search parts were the freely roaming Levy flights. However, as it can be observed that the Levy looping activates the flight paths of the honeybee that aids in conducting the local search around the identified sources of food (Reynolds et al., 2007).

A local search algorithm for the Bees Algorithm was proposed by Hussein et al. (2016) and was referred to as the Greedy Levy-based Local Search Algorithm (GLLSA). The proposed algorithm was based on the Patch-Levy-based Initialization Algorithm which was also proposed by Hussein et al. (2016). Levy looping search is modeled by the GLLSA and was employed for the exploitation of the patches. Patch-Levy model was adopted in the initialization phase of the Bees Algorithm and it improved the global search part of this algorithm. As a result of the enhancement, a variant of the Bees Algorithm was proposed by Hussein et al. (2016) and was known as Patch-Levy-based Bees Algorithm. The proposed algorithm incorporated the integration of the Patch-Levy-based Initialization Algorithm and GLLSA. The Patch-Levy-based Initialization Algorithm was adopted for the initialization stage whereas GLLSA was employed for the local search. The global search was performed in accordance with the Patch-Levy model as incorporated in the Patch-Levy-based Initialization Algorithm.

Moreover, there have been studies that are focused on enhancing the overall performance of the Bees Algorithm. This has been accomplished by hybridizing the

Bee algorithm with the other metaheuristics that leads to improvement in the local and global search parts of the Bees Algorithm (Yuce et al., 2015a; Abdullah and Alzaqebah, 2013; Packianather et al., 2014; Imanguliyev, 2013; Lien and Cheng, 2012; Nguyen et al., 2012; Sadiq and Hamad, 2010).

5.3.3 IMPROVEMENTS MADE IN THE INITIALIZATION OF THE PROBLEM

A strategy for the initialization of the population in the Bees Algorithm was proposed by Imanguliyev (2013). The algorithm enabled the neighborhood search before the inception of the main local search. In the proposed strategy, the scout bees are distributed initially at random as is the process with the basic Bees Algorithm. Then the process of neighborhood search incepts that conducts the search in the region of some of the selected sites. As such the problem associated with increasing number of function evaluation is decreased. The major goal of the proposed strategy was to obtain better information about the surrounding search space. As such the main subsequent local search stage could incept from the patches with higher fitness value. However, an initialization stage wherein the uniform distribution is still adopted which is similar to the other optimization problem. Hence the local search process is mainly the neighborhood search and as such is the initial stage of the algorithm. This is followed with the main search phase of the algorithm. The proposed approach was reported to be advantageous as the strategy entails a different local search strategy than that of the main search process. However, the strategy proposed by Imanguliyev (2013) may not be advantageous as it can be considered to be an early strategy of local search in a previously held iteration. The enhancement on the initial part was not tested alone and as such the proposed improvement couldn't be judged appropriately. The proposed strategy from Imanguliyev (2013) was tested in combination with the adaptive strategy of recruitment. Hence no obvious evaluation of the improved performance could be observed as was expected to be achieved.

An initialization algorithm referred to as Patch-Levy-based initialization algorithm was proposed by Hussein et al. (2014) and was incorporated into the basic Bees Algorithm. The proposed improvement strategy was referred to as Patch-Levy-based initialization Bees Algorithm. The approach proposed was observed to closely mimic the actual behavior of the bees in the biological system. The search space in case of Patch-Levy-based initialization algorithm is divided clearly into transparent segments that are representatives of patches. As such the distributed food sources become easily visible and distributed on the clear observable patches. The bees are then distributed on the basis of Levy flight distribution which are revealed to be approximation of the natural flight patterns of the bees (Hussein et al., 2014).

5.4 CONCLUSION

The present chapter provides an overview of the Bees Algorithm. Various improvements have been discussed. The application of Bees Algorithm has been discussed toward the end of the chapter.

6 Firefly Algorithm

6.1 INTRODUCTION

Firefly algorithm is one of the swarm intelligence methods developed by Yang (2010b) in 2010. Firefly optimization algorithm is stochastic, and metaheuristic algorithm has the potential ability to solve the hardest optimization problem. Being stochastic, it is meant that the optimization algorithm searches for a set of solutions through a kind of randomization process. Flashing light of fireflies is the major source of inspiration for the development of firefly optimization algorithm. In the firefly optimization algorithm, the heuristics conceptual framework concentrates on generating new solutions within the search space of the problem and selects the best solution for survival. On the other hand, the search process is being avoided to get trapped into local optima through the randomization process. A candidate solution is improvised by the local search, and therefore, the solution is placed in local optimum until and unless improvements are detected.

Firefly algorithm is population-based optimization technique. Following advantages are inherited by the population-based algorithms in comparison to the single-point search algorithms (Prugel-Bennett, 2010):

i. Parameter's tuning aids the algorithm to learn good parameter values and hence provide a balance between exploitation and exploration.
ii. The distractions within the search space is avoided through the mechanism of low-pass filtering.
iii. Bad luck in making the initial decisions is hedged.
iv. Focus on searching is dependent on the crossover. This means that the offspring will have the same value of the variable if both the parents share the same value of the variable.

The characteristics of fireflies that led to the development of firefly algorithm have been discussed in the subsequent discussion. Flashing light is one of the main characteristics of fireflies. There are two fundamental functions associated with flashing light: to warn potential predators and to attract mating partners. Few rules of physics are obeyed by the flashing light. The intensity of light I decreases with the distance and is inversely proportional to the square of the distance. This very phenomenon of flashing light inspired by Yang (2010c) and as a result he developed the firefly algorithm. The firefly also acts as an oscillator that charges and discharges the light at regular intervals. A mutual coupling occurs between fireflies when they are placed in vicinity to another flies.

In the next section, biological foundations of the firefly algorithm are discussed. The discussion is followed with a brief discussion on the characteristics of firefly algorithm and culminates into a discussion on the main structure of the firefly algorithm.

6.2   BIOLOGICAL FOUNDATIONS

Fireflies are considered to be the most charismatic of all the insects. There are over 2,000 species of fireflies existing around the world. Fireflies usually live in warm environments and are active during the summer nights. There has been studies on the firefly phenomenon and as a result there exists a number of papers (De Wet et al., 1987; Brasier et al., 1989; Strehler et al., 1952; Deluca, 1962). Bioluminescence is a biochemical process behind the flashing light characterizing the fireflies. Flashing serves to attract the matting partners as well as aids in warning off the potential predators. Some adults in certain species of fireflies are incapable of the aforementioned biochemical process. These species attract their mates due to pheromone.

Fireflies possess light producing organs known as lanterns wherein the bioluminescence reaction takes place. Most of the organisms provide only slow modulated flashes whereas the adults are able to emit high intensity and discrete flashes. The signals emanating from the central nervous system of fireflies are the main source of light producing capability of the lanterns. The firefly species rely on bioluminescent signals. The first signalers are males who attract the flightless females on the ground. The females then emit continuous or discrete flashing lights in response to the male signalers. A distinct flash signal pattern is produced by both the mating partners and is used to encode information like identity of sex and species. The attraction of female fireflies relies on the behavioral differences in the courtship signals. Male signalers that produce brighter signals are usually preferred by the female fireflies. It has been already stated that the intensity of the flashing light is inversely proportional to the square of the distance. Fortunately, some of the female fireflies are not able to distinguish between more distant flashes produced by stronger light source and closer flashes produced by weaker sources of light.

Flash lights produced by fireflies are highly conspicuous and is capable of deterring the potential predators. Flash signals evolve to aid in defense mechanism and hence is relevant to the natural selection process wherein only the strongest individuals can survive.

Self-organization and decentralization are the two main characteristic features of swarm intelligence. The autonomous individuals live in common place as such ants in anthills, bees in hives, etc. However, some communication as well as interactions are required among the members living together. In fact, the members in sociality must adapt to the overall goals within the group in which they live. The social life of fireflies is dedicated to reproduction rather than foraging. Flashing light connects this basic collective decision and hence lays the main biological foundation for the development of firefly algorithm.

6.3 STRUCTURE OF FIREFLY ALGORITHM

The algorithm of firefly is based on physical formula relating light intensity I and the distance r. When the distance from the light source increases, the light becomes weaker, this phenomenon is used for the objective function to be minimized. Following three idealized rules have been considered for describing the firefly algorithm: All the fireflies have been considered to be unisex and hence only firefly will be attracted toward other fireflies irrespective of their sex, secondly, attractiveness has been considered to be proportional to the brightness and hence a firefly with less brighter flashlight will get attracted toward a firefly with brighter intensity flashlight, thirdly, the landscape of the objective function determines and affects the brightness of the firefly. Based on the above rules, structure of firefly algorithm had been formulated and presented as a pseudo code by Yang (2009).

The variation of attractiveness is proportional to the intensity of the light and as such, Equation (6.1) provides for a relationship between attractiveness (β) and distance (r):

$$\beta = \beta_0 e^{-\gamma r^2} \tag{6.1}$$

where, β_0 is the attractiveness for distance $r = 0$.

Equation (6.2) aids in obtaining the movement of a firefly i as it gets attracted to another more attractive firefly j:

$$x_i^{t+1} = x_i^t + \beta_0 e^{-\gamma r_{ij}^2} \left(x_j^t - x_i^t \right) + \alpha_t \in_i^t \tag{6.2}$$

where, second term is significant of attraction. α_t is the randomization parameter and $\in_i^t$ is a vector of random numbers and is drawn from Gaussian distribution. If $\gamma = 0$ then the algorithm reduces to particle swarm optimization (Wolpert and Macready, 1997). However, if $\beta_0 = 0$, then it becomes a simple random walk.

Tuning of algorithm parameters plays a critical role in its successful implementation. α_t is randomness and Equation (6.3) can be used to fine-tune this parameter:

$$\alpha_t = \alpha_0 \delta^t, 0 < \delta < 1 \tag{6.3}$$

where, α_0 is the initial randomness scaling factor and cooling factor is represented by δ which can range between 0.95 and 0.97 (Wolpert and Macready, 1997).

Firefly algorithm will be more efficient in case if α_0 is associated with scaling of the associated design variable. One can set α_0 to $0.01 L$ if L is the average scale of the problem under consideration (Yang, 2009; Das et al., 2011). As far as β_0 is concerned, for most of the optimization problem it can be taken to be 1. On the other hand, γ should be related to scaling L and can be set as $y = 1/\sqrt{L}$. The population size $n = 15$–100 and the best is to range it 25–40 (Wolpert and Macready, 1997).

6.4 CHARACTERISTICS OF FIREFLY ALGORITHM

There are, however, two important issues associated with firefly algorithms: the variation of intensity of the light and formulation of attractiveness. It has been assumed that attractiveness of firefly is determined by its brightness. The brightness is associated with the objective function under consideration.

Brightness I of a firefly at a particular location x can be chosen in accordance with the following: $I(x) \propto f(x)$. Attractiveness β should be left to be chosen by the firefly as its relative and therefore it will vary with the distance r between the two fireflies. The intensity of light, however, decreases with distance from the emanating source. Further it also gets absorbed in the media and therefore attractiveness has been allowed to vary with the degree of absorption. The intensity of light has been revealed to vary in accordance to the inverse square law i.e., $I(r) = I_s/r^2$. In this formulation, I_s is the intensity at the source. Intensity of light, I, varies with the distance r if the medium has fixed light absorption coefficient γ. The singularity for $r = 0$ has been avoided, following Gaussian formulation was given:

$$I(r) = I_0 e^{-\gamma r^2} \tag{6.4}$$

However, for cases where one needs to decipher a function that decreases monotonically at a slower rate, following formulation has been given:

$$I(r) = \frac{I_0}{1 + \gamma r^r} \tag{6.5}$$

For a shorter distance, it can be observed that the formulations given by Equations (6.4) and (6.5) are almost the same.

Furthermore, it has been already stated that the attractiveness β is proportional to the intensity of light and as such Equation (6.6) can be used to define the attractiveness β as follows:

$$\beta(r) = \beta_0 e^{-\gamma r^2} \tag{6.6}$$

$\beta(r)$ can be any monotonically decreasing function and therefore Equation (6.7) represents the generalized formulation:

$$\beta(r) = \beta_0 e^{-\gamma r^m} \tag{6.7}$$

where, $m \geq 1$.

The characteristic length for a fixed light absorption coefficient γ becomes $\Gamma = \gamma^{-1/m} \to 1$ as $m \to \infty$.

Equation (6.8) represents the distance between two fireflies that are at positions x_i and x_j, respectively:

$$r_{ij} = \|x_i - x_j\| = \sqrt{\sum_{k=1}^{d} (x_{i,k} - x_{j,k})^2} \tag{6.8}$$

where, $x_{i,k}$ is the kth component of spatial coordinate x_i of the ith firefly. In 2D case, the distance between the two fireflies can be expressed by Equation (6.9):

$$r_{ij} = \sqrt{\left(x_i - x_j\right)^2 + \left(y_i - y_j\right)^2} \qquad (6.9)$$

Equation (6.10) aids in determination of the movement of a firefly i as it gets attracted to another firefly j:

$$x_i = x_i + \beta_0 e^{-\gamma r_{ij}^2}\left(x_j - x_i\right) + \alpha\left(\text{rand} - \frac{1}{2}\right) \qquad (6.10)$$

The second term represents the attraction, the randomization is reflected in the third term by α and rand generates a random number between 0 and 1.

For most of the cases, β_0 can be taken as 1 and the randomization factor α can range between 0 and 1.

6.5 VARIANTS OF FIREFLY ALGORITHM

Several variants of firefly algorithm have been reported in different literatures. As such a suitable classification scheme is necessary to classify the different variants. The present section throws light on the classification scheme depicted in Fister et al. (2013). Firefly algorithms have been classified in accordance with the following: the modifications made, the manner in which the modifications have been made and the scope of modifications.

From the first aspect, the classifications made are: representation of fireflies, evaluation of fitness function, moving fireflies, population scheme and determination of the best solution. In accordance with the second aspect, following categories of parameter control in firefly algorithm have been considered: self-adaptive, adaptive and deterministic. Modifications in firefly algorithms have been proposed with regard to the third aspect: whole population, the entire firefly and an element of the firefly.

The subsequent sub-section discusses on modified variants as well as hybrid variants of firefly algorithms have been discussed.

6.5.1 MODIFIED VARIANTS OF FIREFLY ALGORITHM

Variation of the light intensity as well as attractiveness forms the ground of firefly algorithm. As such there is lot of scope to improve the firefly algorithm with these two important factors. As for instance, modification of the random movement of brightest firefly in case where current best doesn't improve and the brightness may continue to decrease. The proposed modifications tries to enhance the position of such firefly through generation of m-uniform random numbers. The position of the firefly is made to progress toward its best performance. The brightest firefly will exist to stay in its current best position even if such direction doesn't exists. The brightest firefly in such cases is also considered to be elitist solution as its best position is never replaced by a best-found solution having lower fitness value in the current generation.

A comparative analysis with seven existing benchmark functions revealed that the modified firefly algorithm performed better than the classical one.

Researchers have been able to decipher a number of binary firefly algorithms and have employed them to solve different classes of problems (Palit et al., 2011; Falcon et al., 2011; Chandrasekaran and Simon, 2012). A binary firefly algorithm was proposed by Palit et al. (2011) for cryptanalysis. This binary firefly algorithm aided in determination of plain text from the cipher text through the support of Merkle-Hellman Knapsack algorithm (Forouzan, 2007). New representation of fireflies resulted in changing almost all the components of binary firefly algorithm. A comparative analysis revealed a better performance of the binary firefly in comparison to the genetic algorithm. Falcon et al. (2011) also proposed a binary firefly algorithm that used a binary encoding of the candidate solution. Furthermore for accelerating the search, an adaptive light coefficient was also integrated. A combinatorial optimization problem in the form of system-level diagnostics was solved to carry out the empirical analysis. A comparison analysis of the proposed binary firefly algorithm with the existing particle swarm optimization and artificial immune system revealed a better performing characteristics in terms of convergence speed and memory requirements. A binary coded firefly algorithm was proposed by Chandrasekaran and Simon (2012) utilized it to solve network and reliability constrained unit commitment problem (Carrión and Arroyo, 2006). Ten units of IEEE-RTS (IEEE reliability test system) system were taken to adjudge the effectiveness of the proposed algorithm. The results obtained were revealed to be promising when compared to other reported techniques.

A new firefly algorithm was proposed by Farahani et al. (2011) with due consideration to stabilization of fireflies' movement. The modification resulted in increased convergence speed through the aid of Gaussian distribution that moves all the fireflies to the global best in each iteration. The fixed randomization parameter α was also modified in the new developed firefly algorithm. Five standard functions were used to perform test for the modified firefly algorithm and experimental results revealed better performance and accuracy in comparison to the existing classical firefly algorithm.

Formulation of new firefly algorithm was proposed by Yang (2010c) employing the Levy flights moving strategy. Numerical studies were conducted and the results revealed a superior performance of the proposed Levy flight firefly algorithm over genetic algorithms and particle swarm optimization. The comparison was done for efficiency as well the success rate.

A combination approach integrating chaotic maps and firefly algorithm was proposed by dos Santos Coelho et al. (2011). The integrated algorithm aided in improving the convergence speed of the existing classical firefly algorithm. The chaotic maps aided in easier escape of the solution trapped in local optima. Tuning of the randomized parameter α and absorption coefficient γ was underpinned by the employability of chaotic maps. Power of the proposed firefly algorithm was illustrated through the benchmark test functions. Furthermore, the results obtained were compared to that with the other optimization techniques and revealed that the modified algorithm outperformed the other techniques of optimization. Gandomi et al. (2013) enhanced the global search mobility through the introduction of chaos into firefly algorithm. Attractiveness as well as the absorption coefficients were tuned through the employability of chaotic maps. Influence of using twelve different chaotic maps on the optimization of existing benchmark test functions was analyzed. Analysis revealed that some of the chaotic maps led the chaotic firefly algorithm to outperform the classical firefly algorithm.

A parallelized firefly algorithm was developed by Subotic et al. (2012) and employed it to solve the constrained optimization problem. The proposed algorithm was tested on standard benchmark functions and the tests revealed better results in terms of execution time. However, the conclusion was valid only for more than one population taken into consideration. A modified firefly algorithm was proposed by Husselmann et al. (2012) through a parallel graphical processing unit. The results of the parallel firefly algorithm were revealed to be more accurate and faster in comparison to the original firefly algorithm. However, the presented algorithm was valid for only the multi-modal functions.

6.5.2 Hybrid Variants of Firefly Algorithm

Any two general problem solvers, in accordance with the No-Free-Lunch theorem, are equivalent in performance when their average performance is compared across all the possible problems. That said, the given problem solvers are capable to solve and obtain results on any classes of problems. Specific heuristics are present to solve a specific set of problems and normally improve the results of problem solvers by exploiting the knowledge and concepts specific to the domain of the problem under consideration. Such heuristics can be integrated with firefly algorithms which is a general problem solver. The hybridized firefly algorithm aids in improving the overall results of the problem under consideration. On the other hand, firefly algorithm itself can be hybridized with other problem solvers and yield better results in terms of faster convergence, etc.

A number of works have been reported on the hybridization of firefly algorithm. Eagle Strategy, for instance, was formulated by Yang and Deb (2010) where in the Levy flight search is combined with the firefly algorithm. The eagles have the characteristic features to fly randomly over their territory which is similar to Levy flights (Brown et al., 2007). The eagle tries to catch the prey as and when they see it and in the most efficient manner possible. Eagle Strategy comprises of two important components: random search by Levy flight and an intensive local search mechanism. Firefly algorithm was used to carry out the intensive local search process. The formulated hybrid metaheuristics was applied to the Ackley function with Gaussian noise. The results showed that the proposed algorithm outperformed the particle swarm optimization algorithm in terms of efficiency as well as success rate.

Hybridization of firefly algorithm was discussed by Luthra and Pal (2011) and was employed for cryptanalysis of mono-alphabetic substitution cipher. The algorithm was integrated with operators of mutation and crossover that are used in genetic algorithm. The crossover operator in this case was the gene crossover while for the mutation operator, the permutation mutation was taken into account. It was revealed from the experiments that the algorithm has better performance for large input cipher text lengths. However, a larger number of generations would be required for smaller input cipher lengths.

A hybrid evolutionary firefly algorithm was proposed by Abdullah et al. (2012). The proposed algorithm was a combination of classical firefly algorithm and differential evolution method. The evolutionary operations associated with the differential evolution method was used to enhance the accuracy of the search process as well as to enhance the information sharing among the fireflies. The population of fireflies were divided in accordance with the fitness value. The firefly operator was applied in the first and the evolutionary operators were applied in others. The parameters associated with biological model were estimated using the proposed hybridized algorithm. The experimental results obtained revealed that the hybrid algorithm performed far better than the particle swarm optimization, genetic algorithms, classical firefly algorithm and evolutionary programming.

In another study, the classical firefly algorithm was hybridized using a local search heuristic (Fister et al., 2012). The proposed hybridized approach was then adopted to solve graph-3-coloring optimization problem (Fister et al., 2013). The results obtained were compared with that of hybrid evolutionary algorithm and evolutionary algorithm with SAW (Eiben et al., 1998). Obtained results showed encouraging results and demonstrated that the proposed algorithm could be applied to address the combinatorial optimization problems.

Firefly algorithm was utilized in a study for training the parameters of fuzzy neural network (Hertz and de Werra, 1987) and the proposed algorithm was used to recognize speech. It was observed that the firefly algorithm could improvise the ability of generalizing fuzzy neural networks. The obtained results showed that the proposed hybridized approach had a faster speech recognition rate in comparison to the classical fuzzy neural network approach trained by the particle swarm optimization method. In another study (Nandy et al., 2012) firefly algorithm was utilized to train feed-forward neural network. The firefly algorithm was incorporated into back propagation algorithm and aided the feed-forward neural network to achieve faster convergence rate. The proposed algorithm was tested over standard sets of data and it was observed that the convergence to local optima was achieved in few iterations only.

Cellular learning automata was hybridized with the classical firefly algorithm in a study conducted by Hassanzadeh and Meybodi (2012). The cellular learning automata was responsible for diversified solutions in the population of firefly and firefly algorithm was incorporated to improve these solution. Five well-known benchmark functions were used to adjudge the performance of the proposed algorithm. The results revealed that the algorithm was able to find global optima and enhance the exploration rate of standard firefly algorithm.

Three classes of algorithms were proposed by Farahani with due consideration to improve the performance of the firefly algorithm. Learning automata was used in the first class and was employed to adapt the absorption and the randomization parameters. In the second class, genetic algorithm was integrated with firefly algorithm and hence aided in balancing the exploration as well as the exploitation properties of the aforementioned new proposed metaheuristics in the last class, random walks based on the Gaussian distribution was used which helped the fireflies to move over the search space. Experimental tests were performed on the benchmark functions and the competitiveness of the proposed algorithms over the classical firefly algorithm was revealed. The proposed algorithm also was more competitive in comparison to the particle swarm optimization.

A Flexible Neural Tree model for micro-array data was developed by Aruchamy et al. and was used to predict the cancer through the utilization of ant colony optimization (ACO). Firefly algorithm was employed to fine-tune the parameters associated with neural tree. The modified algorithm aided in finding the optimal solution at a faster rate of convergence and lower error. A comparative analysis was performed between firefly algorithm and the exponential particle swarm optimization technique.

6.6 ENGINEERING APPLICATIONS OF FIREFLY ALGORITHM

Firefly algorithm has been used widely used for solving problems of engineering practice. There are numerous engineering applications and various researchers have used the same in problems related to wireless sensor network, chemistry, meteorology, semantic web, civil engineering, robotics, antenna design, image processing, etc. and hence it can be observed that most of the applications have been in the domain of industrial operations.

6.7 CONCLUSION

Present chapter has reviewed the fundamentals associated with firefly algorithm. Firefly algorithm has gained its stature and is now widely adopted for a large number of engineering applications. The development in this area has been dynamic as newer applications appear almost every day. The credibility of firefly algorithm to address more challenging problems in the near future cannot be undermined. The importance of exploitation as well as the exploration phases has also been highlighted in the chapter. Various modifications have also been discussed along with the hybridized variants of firefly algorithms.

7 Cuckoo Search Algorithm

7.1 INTRODUCTION

Cuckoo search is an evolutionary optimization technique that was developed by Yang and Deb (2009). The associated theory for the development of cuckoo search algorithm is based on the species of the bird cuckoo. Cuckoos have been known for their sounds as well as their aggressive strategy of reproduction. Through their reproduction strategy, the mature cuckoo birds lay their eggs in the nests of other host birds. This mechanism is referred to as obligate brood parasitism.

The solution is represented by each of the eggs present in the nest. A new solution is represented by the cuckoo. The basis of the cuckoo algorithm lies in the specific egg laying and breeding of cuckoos. In this case, the new eggs are either thrown away by the host bird if it is able to discover the alien eggs. In some cases, the host bird may abandon the nest and then builds a new nest at some other location.

In general, the eggs laid by cuckoo hatches earlier than the eggs of the host birds. Some of the eggs may also have a similar appearance as compared to the host egg birds and can also mimic them. These eggs have an opportunity to grow up and receive comparatively more feedings than their host counterpart. As such, these eggs will transform into mature cuckoos if such eggs are not recognized by the host bird.

In the cuckoo search algorithm, two forms exists: the mature cuckoo and their eggs. A society is developed once the remaining eggs grow and turn into mature cuckoos. The features of the environment as well as the immigration of such societies of cuckoos results into their convergence and hence identify the best environment required for breeding and hence reproduction. Therefore, the nest solution to an objective problem is found through the best environment required for efficient breeding and reproduction activities.

The present chapter provides a brief overview of the cuckoo search algorithm. The applications addressed are also presented. The chapter finally terminates with the concluding remarks.

7.2 CUCKOO SEARCH METHODOLOGY

There are three basic rules for the basis of cuckoo search algorithm: Each cuckoo bird lays only one egg at a time and dumps the laid egg in random chosen nest, the best nest consisting of the best quality eggs is carried forward to the next generation and there is a fixed number of available hosts' nest and the egg laid by the cuckoo in the hosts' bird nest is discovered with a probability of p_a that ranges in the interval 0 and 1.

There are two possibilities that arises in a follow-up to the three aforementioned rules: The host bird either deciphers the new egg or it abandons the nest and builds a new nest. The last assumption can be approximated by probability parameter p_a.

The optimization problem can be solved if the values of the problem variables can be formed as an array. In the cuckoo search algorithm, this array is referred to as habitat. Therefore the following are employed in order to obtain the cuckoo habitat: A habitat is represented as: Habitat $= [x_1, x_2 \ldots x_{Nvar}]$ and can be represented by $1 \times N_{var}$ and is significant of the current living position of cuckoo.

Profit function is represented as Profit $= f_p$(habitat) $= f_p(x_1, x_2 \ldots x_{Nvar})$. The profit of the identified habitat is obtained through the calculation of the profit function. If there is a cost function, then Profit $= -$Cost(habitat) $= f_c(x_1, x_2 \ldots x_{Nvar})$. Therefore the cuckoo search algorithm is for maximization of profit function.

One of the important considerations in the cuckoo search algorithm is the radius in which the eggs are laid and this is referred to as ELR. This is also known as maximum distance i.e., the maximum distance from the habitat in which the cuckoos lays their eggs. Each cuckoo has its specific ELR in an optimization problem that has an upper (var_{hi}) and lower (var_{low}) limits on their variables. The ELR is proportional to the total number of eggs variable limits and number of current eggs of cuckoos. Therefore ELR is defined by the following equation:

$$\text{ELR} = \alpha \times \frac{\text{Number of current eggs of cukoo}}{\text{Total number of eggs}} \times (var_{hi} - var_{low}) \qquad (7.1)$$

where, α is an integer and is meant to handle the maximum value of ELR.

The discussion now ensues on the cuckoos' style of laying eggs. Each cuckoo starts to lay eggs in other host bird's nest randomly. The eggs are laid in the nest that are within the ELR. Once the eggs are laid, the host bird identifies the less similar eggs and then throws the detected non-similar eggs out of its nest. Hence after the egg laying process is over, $p\%$ of all the eggs, having less profit values will be thrown out of the nests. The thrown out eggs then have no chances to grow. The remaining eggs are hatched and fed by the host bird.

Also, there is only egg that has the chance to grow in the host bird nest. This is because, as the cuckoo egg hatches and becomes a chick it will throw out the other eggs in the nest. In case, the host bird's egg hatches first in comparison to the cuckoo egg, the cuckoo chick will eat most of the food brought by the host bird. As a result, the cuckoo chick survives and the host bird's chick dies of hunger.

After egg laying process has been completed by the cuckoo, immigration is the next step in the cuckoo search algorithm. Young cuckoos grow and become mature and for some time will live in the society. However, with the passage of time and as the egg laying process approaches again, the cuckoos immigrate to a better habitat surroundings. The immigration to the new area is such that their eggs are similar to the other host birds in the new habitat. Also, the availability of food for young cuckoos is more. Thus a cuckoo group will form in a new area. The society that associates itself with the best possible profit forms the goal point for the cuckoos to immigrate.

When the mature cuckoos live all over the environment, it becomes difficult to differentiate the group to which the cuckoos belong. Therefore grouping of cuckoos becomes essential and is accomplished by means of k-means clustering method. Clustering method involves the grouping of cuckoos into the clusters and finding the best group and therefore ultimately aids in selection of the goal habitat. In order to arrive at the best group, the profit value is determined for each of the cuckoo groups. The goal group is then determined by maximum value of the calculated mean profit value. The habitat corresponding to the best group then becomes the new destination habitat for the immigration cuckoos. The cuckoos do not completely fly to the destination habitat, but they travel only a part of the way and are accompanied with a bit of deviation. Therefore each cuckoo, as for instance, flies for $\lambda\%$ of the total distance and with an associated deviation of φ radians. The two parameters will aid the cuckoos to identify more positions in the environments. The two parameters for each cuckoo are defined as follows:

$$\lambda \sim U(0,1)\, \text{and}\, \varphi \sim U(-\omega,\omega) \tag{7.2}$$

where, $\lambda \sim U(0,1)$ means that λ is a random number ranging 0 and 1 and ω is a parameter that aids in constraining the deviation and for a good convergence, a value of $\pi/6$ radians is considered to be good enough. Each matured cuckoo is given some eggs when all the cuckoos have migrated themselves to the goal habitat. An ELR is then calculated for each cuckoo by taking into consideration the number of eggs dedicated to each bird. The process of laying eggs starts again.

There is, however, a limit to the maximum number of cuckoos that can live in a cuckoo society. As such, only N_{max} number of cuckoos having better profit values will survive and the rest will die. The limitation occurs because of the limited food resources, the inability to identify suitable nests for the eggs and being killed by the predators.

For the last process, the entire cuckoo population will migrate toward the best habitat wherein there is maximum similarity to the eggs of host bird and also the maximum availability of food sources. There will be fewest loss of eggs in the identified one best habitat. The identified habitat will produce maximum profit. The steps of the cuckoo algorithm are presented as below:

Step 1: Generate initial population of n number of host nests. The eggs in the host nests are the candidates for optimal parameters and represented as (a_k, r_k).

Step 2: The process of egg laying is initiated by cuckoo. The eggs laid by cuckoo in the kth nest is represented as (ak', bk'). The kth nest is selected randomly by cuckoo. The eggs laid by cuckoo is almost similar to that of the eggs in the host birds' nest.

Step 3: The fitness of eggs laid by cuckoo is compared with the fitness of the host birds' eggs.

Step 4: If the calculated fitness value of the cuckoos' eggs is better than that of the host birds' eggs. Replace the eggs in the nest by the cuckoos' eggs.

Step 5: If the eggs are noticed by the host bird, then the nest is abandoned by the host bird and a new nest is built.

Steps 1–5 are iterated until and unless the termination criterion is achieved.

One of the important issues in the cuckoo search algorithm is the role of Lévy flights in the generation of new solution, $x^{(t+1)}$, where $x^{(t+1)} = x_i + sE_4$. Here in the equation, E_4 is drawn from a standard normal distribution or drawn from Lévy distribution for Lévy flights. The similarity between the host's egg and a cuckoo's egg aids in providing a link for the random walks. However, this may be tricky in implementation. The step size s determines the distance that can be traversed by a random walker for a fixed number of iterations.

The generated new solution will be far away from the old solution in case the value of step size is very large. As such, the move is unlikely to be accepted. If the value of step size is pretty small, then the change is too small to be significant and hence such a search process is also not considered to be an efficient approach. Therefore, a suitable step size is essential for maintaining an efficient search process.

7.3   VARIANTS OF CUCKOO SEARCH ALGORITHM

7.3.1   Adaptive Cuckoo Search Algorithm

The adaptive cuckoo search algorithm was proposed by Pauline et al. and was used to address the structural engineering problems. The structural engineering problem encompasses safety, reliability, stability, cost of production, rigidity and sustainability aspects of building structures. Few of the aforementioned issues have been tried to be fixed through the implementation of adaptive cuckoo search algorithm. The proposed algorithm employs strategy of adaptive step size selection and the process of diversification (Pauline et al., 2017). The proposed approach therefore maintains a balance between the diversification and intensification performances of the cuckoo search algorithm. The effectiveness of the proposed algorithm was demonstrated through the three structural engineering problems: design of gear train, three-bar truss and pressure vessel design. The demonstration revealed the effectiveness of adaptive cuckoo search algorithm. The validity was adjudged through the test in benchmark functions.

7.3.2   Self-Adaptive Cuckoo Search Algorithm

Camera calibration is one of the most important process in computer vision. In a view to calibrate the optimization process, a self-adaptive cuckoo search algorithm was proposed by Liu and Qi (2016). A step length formula was proposed to adaptively adjust the step size and hence avoid the algorithm being trapped in local minima and therefore miss the global best solution. It was depicted that the initial estimation values are required by the improved version for cuckoo search algorithm and is used in a nonlinear camera. The proposed algorithm was reported to solve traditional optimization problems, in particular the problems sensitive to the initial values. The algorithm integrates itself with the process of camera calibration. Through such integrative approach, it was possible to cumulate the intrinsic parameters of camera and hence the coefficient of radial distortion. The analysis for the projection error, standard deviation and mean absolute error was done for the different noise levels. The experimental evaluation was demonstrated and the proposed algorithm was revealed to be more accurate and robust.

7.3.3   Cuckoo Search Clustering Algorithm

A novel algorithm was proposed by Zhao et al. (2016) that was inspired by the importance of protein complexes in understanding the activity of cell machinery.

The proposed algorithm was referred to as cuckoo search clustering algorithm. The proposed algorithm was able to detect the protein complexes easily in comparison to the other detection algorithms. A dynamic protein network was constructed at first and the proteins were then detected in each of the constructed complex cores. Subsequently the protein attachment has been clustered with the cuckoo search process. The proposed algorithm was found to be suitable in comparison to the fact finding results based on (Database of Interacting Proteins) DIP datasets and Krogan dataset.

7.3.4 NOVEL ADAPTIVE CUCKOO SEARCH ALGORITHM

In order to enhance the contrast associated with the satellite images, Suresh et al. modified the cuckoo search algorithm and was known as the novel adaptive cuckoo search algorithm. The proposed algorithm was employed for enhancing the visual contrast of the satellite images wherein the main focus is on the contrast improvement of the procured images and naturalness conservation to the maximum possible. The proposed algorithm was implemented in three phases: In the first phase, a chaotic initiation was employed which aided in avoidance of the premature convergence of the images. The second phase emphasizes the importance of the convergence rate of the cuckoo search algorithm. This has been improved greatly through the aid of fitness value which in turn depends on the strategy of adaptive Lévy flight strategy. A strategy of mutation randomization has been incorporated in the final phase that helps in facilitating the balance between exploitation and exploration phases. This ultimately ensures the best possible optimum solution. The proposed algorithm was proved to give better results with the adopted quantitative evaluations.

7.3.5 CUCKOO SEARCH ALGORITHM BASED ON SELF-LEARNING CRITERIA

Cuckoo search algorithm based on self-learning criteria was proposed by Nguyen and Fujita. In the proposed variant for the cuckoo search, the optimization of the learning stage was combined and thereby enhancing the overall performance of the cuckoo eggs (Nguyen and Fujita, 2016). A hybrid factor was thus incorporated to avoid the cuckoo eggs getting stuck in the local optima. The proposed modified cuckoo algorithm was adopted for two IEEE standard systems to prove its effectiveness: 30 bus systems and the 57 bus systems. The proposed algorithm when implemented to the 30 bus system aided in the evaluation of the 3 types of fuel cost functions: multi-fuel cost function, value-point effect cost function and the normal quadratic function. On the other hand, in case of the 50 bus system, the evaluation was done for the different types of sources for the shunt reactive power. The results revealed through the study indicated that the method could ensure better solutions over the other algorithms.

7.3.6 DISCRETE CUCKOO SEARCH ALGORITHM

The discrete cuckoo search algorithm was proposed by Ouaarab and Ahiod. The main source of the proposed algorithm was the occurrence of expensive and unrealistic computations with the increasing number of combinations with respect to the size of the optimization problem. The proposed algorithm was employed to solve the traveling salesman problem (Ouaarab and Ahiod, 2014) wherein a salesman is required to travel finite number of cities only once and the minimization is to be done with respect to the cost of the trip and the distance traveled. It was also suggested that the combinatorial problems could be addressed by extending and reconstructing the population of the proposed algorithm.

A discrete search cuckoo search optimization algorithm was also adopted to address the scheduling problem associated with the cyclic robotic cell (Majumder and Laha, 2016). The problem of scheduling cyclic robotic cell can be referred to as a two-machine robotic cell scheduling for one-unit cycle along with the sequence dependent setup time. For each parts, there will be unique loading and unloading patterns. The cycle time was minimized by determining the robotic moves as well as the sequence of the parts. A fractional scaling factor was proposed to address the problem. The scaling factor was based on step length of Lévy flight distribution over the discrete values. A response surface methodology was used to enhance the execution speed of the proposed algorithm.

7.3.7 DIFFERENTIAL EVOLUTION AND CUCKOO SEARCH ALGORITHM

Elahi et al. proposed a scheme integrating differential evolution with the cuckoo search algorithm. The proposed algorithm was used to address one of the major hurdles in orthogonal frequency division multiplexing. This major hurdle is that of sidelobes of the subcarriers. The modified cuckoo search algorithm was based on the conceptual framework of carrier cancellation. The authors tried to insert a number of cancellation carrier over the edges of the original signal. The integrated optimization framework is then used to obtain the associated amplitudes. A significant sidelobe reduction was revealed through the simulation results.

7.3.8 CUCKOO INSPIRED FAST SEARCH

Ismail et al. proposed a cuckoo inspired fast search algorithm and employed it for fractal image compression (Ismail et al., 2018). The proposed algorithm addressed the major constraint in the technique of image compression. An ordered vector of range of blocks being considered through their coordinate distance as well as through their similarity was taken into consideration. The robustness of the results obtained was evinced through the experimental results. The cuckoo inspired fast search was also revealed to be a scalable technique and also there was a wide reduction in the calculated mean square error. This is because there are only four transformations of dihedral group and is therefore sufficient for making a comparative analysis.

7.3.9 CUCKOO SEARCH ALGORITHM INTEGRATED WITH MEMBRANE COMMUNICATION MECHANISM

A precise dynamic model is required for carrying out design and analysis of overhead crane system. The objective was achieved through a novel radial basis function neural network which was presented for this purpose by Zhu and Wang (2017). One of the biggest challenge in the radial basis function neural network is the determination of reasonable parameters. Therefore cuckoo search algorithm was employed in combination with the membrane communication mechanism to optimize the associated parameter. The population diversity was also preserved through the employability of communication mechanism. An improved accuracy of the algorithm was also ensured through chaotic local search strategy. The benchmark functions were also tested upon by the proposed algorithm and revealed its robustness. The effect of communication set size was also analyzed. The modified cuckoo search algorithm was then used with the radial basis function neural network to design the overhead crane system. The efficiency as well as the effectiveness of the proposed algorithm was therefore demonstrated.

7.3.10 Master-Leader-Slave Cuckoo

A wide range of problems can be addressed using artificial neural networks (ANNs) and therefore to achieve the best possible solution for any problem, selection of a suitable ANN model is very critical. In a study conducted by Jaddi et al., an optimization strategy was proposed to select a suitable ANN model. The proposed optimization algorithm was based on the cuckoo search algorithm. Master-leader-slave multi-population was employed and hence a modified cuckoo search algorithm was developed. Under the influence of master-leader-slave mechanism, some slaves exhibit the best fitness function and therefore will be opted by the leader. The ability of the proposed algorithm was adjudged through the test on benchmark functions.

7.3.11 Cuckoo Search Algorithm with Wavelet Neural Network Model

A vital role in the power system is played by electricity forecasting. However, most of the existing models are single forecasting indicator as for instance short-term electricity price forecasting, short-term load forecasting and short-term wind speed forecasting model. As such, Xiao et al. proposed a new model that was based on singular spectrum analysis and modified wavelet neural network. A modified cuckoo search algorithm was proposed for the evaluation of weights that are initially considered and also for the identification of attributes of dilation and attributes in the wavelet neural network. Several case studies related to the half-hourly electrical price. The experimental results revealed robust performance of the proposed algorithm over the other short-term models.

7.4 ENGINEERING APPLICATIONS OF CUCKOO SEARCH

There are numerous applications of cuckoo search in the domain of engineering. The present section throws light on such applications. Yang and Deb (2010) minimized the weight of the spring, fabrication cost and maximum end deflection through the utilization of different optimization techniques: cuckoo search, genetic algorithm (GA), Lévy flights and particle swarm optimization (PSO). The process parameters involved were diameter of the wire, mean coil diameter, width, length of the welded area and thickness of the beam. The main purpose of the study was to develop cuckoo search algorithm to solve engineering design optimization problem. Noghrehabadi et al. (2011) employed a number of optimization techniques such as cuckoo search algorithm, hybrid power series integrated with cuckoo search and Lévy flights to solve the problem of buckling and deflection of fixed-fixed microbeam-type actuators. The results of the applied algorithm were compared and revealed that the proposed hybrid power series integrated with cuckoo search with eight terms is in good agreement with the numerical results. The problem of optimum design of steel and frames was addressed by Kaveh and Bakhshpoori (2013) through numerous optimization techniques: Standard Cuckoo, Lévy Flights, Genetic Algorithm (GA), Ant Colony Optimization (ACO), Improved Ant Colony Optimization (IACO), Hybrid Big Bang-Big Crunch HBB-BC, Imperialist Competitive Algorithm (ICA), Mantegna's Algorithm, Harmony Search (HS) and Particle Swarm Optimization (PSO). The process parameters involved were elasticity, yield stress and load. Yang and Deb (2013) employed a number of optimization techniques: Standard Cuckoo, Lévy Distribution, Vector Evaluated Genetic Algorithm (VEGA), Pareto Front, Multiobjective Differential Evolution (MODE), Differential Evolution for Multiobjective Optimization (DEMO), Strength Pareto Evolutionary Algorithm (SPEA), Multiobjective Cuckoo Search (MOCS), Non-dominated Sorting Genetic Algorithm (NSGA-II) and Multiobjective bees algorithm (MOBA) and minimized the end deflection, fabrication cost, overall mass and braking time. The results revealed that MOCS performed better in comparison to other algorithms. A number of optimization problems: design problem of a pin-jointed plane frame with fixed base, minimize the weight of corrugated bulkheads for tanker, minimization of vertical deflection of I-beam, minimization of the volume of oil, minimization of the cost of the gear ratio of the gear train, optimization the total cost of a reinforced concrete beam, minimization of the total weight of the speed reducer, pressure vessel design, parameter identification of structures were addressed using a number of optimization techniques: Standard Cuckoo, Lévy Flights, Adaptive Response Surface Method (ARSM), PSO, Hybrid Particle Swarm Optimization (HPSO), GA, Improved ARSM, Differential Evolution (DE) and HPSO with Q learning (Gandomi et al., 2013).

Cuckoo search algorithm has been applied to optimize a number of non-traditional machining processes. Sohrabpoor et al. (2016) used cuckoo search algorithm for multiple objective optimization of electrochemical process. The different process parameters considered were: concentration of the electrolyte, electrolyte flow rate, applied voltage and feed rate of the cemented tungsten carbide tool material. The surface responses considered were material removal rate, radial overcut and surface roughness. Cuckoo search optimization technique has also been used by Madic et al. (2015) for carrying out multiple objective optimization for the laser beam machining process. Laser power, assist gas pressure, cutting speed and focus position were the considered process parameters. Heat affected zone, surface roughness and top kerf width were the response parameters. Stainless steel was the tool material. Mohamad et al. (2015) carried out single-objective optimization of hybrid machining process, wherein the surface roughness of the workpiece material was optimized. Al7075-T6 wrought alloy was the tool material. The process parameters considered during the optimization process were: traverse speed, standoff distance, water jet pressure at nozzle, flow rate of abrasive particles and the abrasive grit size. Cuckoo search algorithm was used for the multiple objective optimization of electrochemical process (Teimouri and Shorabpoor, 2013). Cemented carbide was the tool material. Material removal rate and surface roughness were the response parameters.

7.5 CONCLUSION

The present chapter delineates an overview of the cuckoo search algorithm. Chapter initiated with description of the basic conceptual framework of the standard cuckoo search algorithm. Different variants of the cuckoo search algorithm were discussed. Finally some discussion on the engineering applications of cuckoo search was made. Cuckoo search algorithm has been revealed to be the very efficient optimization algorithm and the robustness has been adjudged through tests on benchmark functions.

Section III

Application of Heuristic Techniques Toward Engineering Problems

8 Engineering Problem Optimized Using Genetic Algorithm

8.1 INTRODUCTION

Ultrasonic machining process was developed in 1950s and was first employed for the finishing operation. Ultrasonic machining process has the potential ability to machine non-conductive and brittle materials that are hard to be machined using any of the conventional machining methods and also processes such as electrochemical machining and electro discharge machining. (Jadoun et al., 2006). Ultrasonic machining machines advanced materials that have hardness greater than HRc 60. The automotive applications of ultrasonic machining process involves: ball bearings, drawing molds, heat exchangers, electronic parts and other minor parts. One of the major advantages of the ultrasonic machining process is that there is no thermal damage to the workpiece material and hence doesn't induces significant levels of residuals stresses. This advantage is very critical for the survival of the brittle material during its service period (Thoe et al., 1998). However, lower rates of production, higher rate of tool wear and low surface roughness (SR) have been reported with application of ultrasonic machining process to the tough materials. However, in case of the brittle materials the different outcomes are reversed i.e., higher rates of production, higher SR and lower rate of tool wear. Plasticity of the tough materials forms the basis of material removal process and a greater tendency of production of surfaces with texture has been reported with denser and non-porous materials (Dam et al., 1995).

Increase in material rate for the ultrasonic machining process is greatly dependent on the energy imparted to the workpiece material in terms of the abrasive grit size, static load and amplitude of the tool tip. A higher impact force can result in cracking of the cone and subsequently the crater damage (Rajurkar et al., 1999). Hence an optimized ultrasonic machining process is very critical to obtain the products with least defects and higher geometrical accuracy. The present chapter delineates the application of genetic algorithm to minimize the SR of a hole as well as maximization of the material removal rate (MRR). A number of parameters such as concentration of the abrasive slurry, feed rate of the tool, power rating and abrasive grit size have been considered.

The chapter incepts with the basic operating mechanism of genetic algorithm through an example. Then the genetic algorithm has been depicted for the optimization of the ultrasonic machining operation carried out on the hole surface.

8.2   DETAILS OF ULTRASONIC MACHINING PROCESS

The working of ultrasonic machining process involves the conversion of higher frequency electrical energy to mechanical vibrations. The conversion takes place via transducer/booster combination. The converted energy is transmitted to energy focusing as well as amplifying device: horn/tool assembly. Vibration of the tool takes place at ultrasonic frequency with an amplitude ranging 12–50 µm. A feed force is required to press the tool. The machining zone is flooded with the abrasive slurry that is generally in the form of water-based slurry. The tool material vibrates over the workpiece and the indenting operation is performed by the abrasive particles that indents the workpiece material. Material removal process is therefore achieved through the crack initiation and propagation mechanism as a result of brittle fracture. The power rating ranges 100–1,000 W. Abrasive slurry in the form of silicon carbide, aluminum oxide and boron carbide suspended in certain medium is pumped in the gap between the tool and the workpiece material. The tool is pushed on the workpiece with some static load. The abrasive particles are continuously indented by the hammer with the kinetic energy imparted by the vibrating tool. Continuous flushing of the abrasive slurry, refreshes the abrasive particles in the machining zone and at the same time also removes the debris from the machining area.

8.3   DETAILS OF THE EXPERIMENTATION PROCESS

The experiments were conducted on ultrasonic machine (USM) with frequency of vibration to be about 20 kHz. Design of experiments (DOE) has been used to plan the experiments and therefore the effect of process parameters on the responses of USM were studied. DOE is an efficient process that aids in suitable planning of the experiments and the number of factors that are required to be investigated. The machining characteristics of the USM process was evaluated by employing a flat workpiece made of zirconia and boron carbide powder is used as an abrasive slurry. The boron carbide powder had different grit sizes and was mixed in water. A circular stainless steel tool was used and was silver-brazed to the tip of the tool. Response surface methodology (RSM) was used to establish the mathematical correlation between the process parameters and the different responses. Central composite design scheme was adopted that ran 31 experiments with a value of alpha equivalent to 2. The experiments carried out were in accordance with the seven center points. Four distinct factors, each with five different levels were considered to run the experiments. Pilot experiments were conducted to obtain the range values for each of the factors. The average diameter of the slurry ranged between 26 and 64 µm, concentration of the slurry was varied between 30 and 50 g/L, rating of power supply ranging 300–500 W and the rate of tool feed was ranged between 0.84 and 1.32 mm/min.

The MRR during the machining operation was obtained using the following equation:

$$\text{MRR} = \frac{W_1 - W_2 - w}{t} \tag{8.1}$$

where W_1 and W_2 are the weights of the workpiece material and that of the workpiece. w is the weight of cut-out portion from the workpiece, time duration of the machining process is represented by t. The SR values were obtained from the entry to the exit of the machined hole.

8.4   DEVELOPMENT OF EMPIRICAL MODELS BY USING RESPONSE SURFACE METHODOLOGY

Response surface methodology has been employed in order to establish relationship between machining parameters and responses. A general form of second-order polynomial response surface empirical model is presented as follows and has been used to arrive at the empirical models:

$$Y_u = \beta_0 + \sum_{i=1}^{n} \beta_i X_{iu} + \sum_{i=1}^{n} \beta_{ii} X_{iu}^2 + \sum_{i=1}^{n} \beta_{ij} X_{iu} X_{iu} + e_u \qquad (8.2)$$

where, corresponding response is represented by Y_u and X_{iu} is the coded value of the ith parameter of machining, the regression coefficients are represented by β_0, β_i, β_{ii}, β_{ij} and the experimental error e^u represents the experimental error of the uth observation. In the present study, abrasive grit size (X_1), concentration of the slurry (X_2), power rating (X_3) and feed rate of the tool (X_4) have been considered as the process parameters. The obtained experimental results have been depicted in Table 8.1. Table 8.1 delineates the obtained results with the different combination of the process parameters considered. RSM has been used then to obtain the empirical relationships for MRR and SR. The obtained equations have been presented as follows:

$$Y_{MRR} = 0.030547 - 0.000032X_1 - 0.000814X_2 + 0.000196X_3 + 0.077875X_4$$

$$+ 0.000014X_1^2 + 0.000021X_2^2 + 0.011659X_4^2 + 0.000019X_1X_2$$

$$+ 0.000002X_1X_3 - 0.000511X_1X_4 + 0.000002X_2X_3$$

$$- 0.002247X_2X_4 - 0.000070X_3X_4$$

$$Y_{SR} = 0.066781 - 0.03638X_1 - 0.00779X_2 + 0.00064X_3 + 0.40698X_4$$

$$+ 0.00074X_1^2 - 0.00003X_2^2 + 0.19711X_4^2 + 0.00076X_1X_2$$

$$- 0.00233X_1X_4 - 0.000001X_2X_3 + 0.00574X_2X_4 - 0.00302X_3X_4$$

8.5   OPTIMIZATION USING GENETIC ALGORITHM

An appropriate selection of the process parameters should be able to enhance the quality of different responses under consideration. For the present study the responses are MRR and SR. For the present case, genetic algorithm has been employed to

TABLE 8.1
Design of Experiments and Observed Responses

Exp. No.	Process Parameters with Un-Coded Value				Responses	
	Feed Rate (mm/min)	Power Rating (W)	Slurry Concentration (g/L)	Grit Size (μm)	Roughness Average (Ra) (μm)	MRR (g/min)
1	0.82	400	35	32	0.72	0.1405
2	1.18	350	40	22	0.56	0.1236
3	1.06	400	35	14	0.65	0.1246
4	0.94	350	30	42	0.91	0.1592
5	0.94	350	30	22	0.55	0.1215
6	0.94	450	40	22	0.53	0.1218
7	0.94	450	30	22	0.58	0.1208
8	1.06	400	35	32	0.65	0.1377
9	0.94	350	40	42	0.99	0.1583
10	1.06	400	25	32	0.62	0.1403
11	1.18	350	30	22	0.55	0.1254
12	1.06	500	35	32	0.72	0.1383
13	1.18	350	30	42	0.91	0.1572
14	1.18	350	40	42	1.03	0.1535
15	1.18	450	30	22	0.56	0.1272
16	0.94	350	40	22	0.52	0.1202
17	1.18	450	40	42	1.03	0.1550
18	1.06	400	35	32	0.67	0.1373
19	1.06	400	35	32	0.65	0.1380
20	1.06	400	35	32	0.67	0.1380
21	1.06	400	45	32	0.74	0.1386
22	0.94	450	40	42	0.99	0.1604
23	1.06	400	35	61	1.54	0.1926
24	0.94	450	30	42	0.93	0.1597
25	1.06	400	35	32	0.67	0.1376
26	1.30	400	35	32	0.76	0.1399
27	1.18	450	30	42	0.89	0.1540
28	1.18	450	40	22	0.56	0.1239
29	1.06	400	35	32	0.69	0.1373
30	1.06	300	35	32	0.73	0.1347
31	1.06	400	35	32	0.67	0.1355

optimize the responses and hence achieve the optimal parameter setting. A general form of optimization process for a simple genetic algorithm has been depicted below:

```
functionsga ()
    {
        Initialize population;
        Calculate fitness function;
        While (fitness value! = termination criteria)
```

```
    {
    Selection;
    Crossover;
    Mutation;
    Calculate fitness function;
    }
}
```

It is the selection operator that determines on as to what solutions are required to be preserved and what all are allowed to die out. However the selection operator maintains the size of the population while segregating between bad and good solutions. There are number of selection operators: tournament selection, roulette wheel selection, proportionate selection, rank selection, steady state selection, etc.

In case of tournament selection, few individuals play several tournaments. The winner of each tournament is selected for the next generation. Figure 8.1 illustrates the process of tournament selection:

The crossover operator aids in creation of new solution from the solutions that are existing currently. The exchange of gene information takes place between the solutions in the mating pool.

Mutation mechanism, on the other hand, results in to introducing new features into the solution strings and as such aids in maintenance of diversity.

The function is evaluated through the fitness value. Optimality of a solution is ensured through the fitness value. As such the fitness value is employed to rank a particular solution over the other solutions. As for instance, consider the following minimization problem:

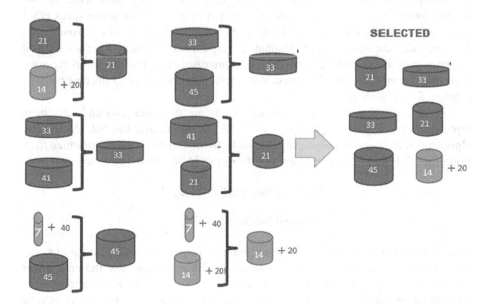

FIGURE 8.1 Tournament selection process.

Minimize $f(d,h)=c((\pi d^2/2)+\pi dh)$,
Subject to:
$g_1(d,h)=(\pi d^2 h/4)\geq 300$
$d_{min}\leq d \leq d_{max}$,
$h_{min}\leq h \leq h_{max}$
$c = 0.0654$

let $d = 8$ and $h = 10$, then $f(s) = 23$

The optimization problem under consideration requires an objective function to be minimized under the influence of constraint functions. For ultrasonic machining, optimization problem can be defined as below:

- Identification of optimum process parameters: X_1, X_2, X_3 and X_4
- Minimization of SR
- Maximization of MRR
- The constraints are as follows:
  - 14 μm ≤ abrasive grit size (X_1) ≤ 65 μm
  - 25 g/L ≤ concentration of abrasive (X_2) ≤ 50 g/L
  - 300 W ≤ power rating (X_3) ≤ 550 W
  - 0.80 mm/min ≤ feed rate of tool ≤ 1.35 mm/min

Genetic algorithm simulates the biological behavior of biological evolution process based on the Darwin's theory of survival of the fittest. A set of potential solutions (chromosomes) initializes the process of genetic algorithm optimization process. These solution are selected randomly and forms the population. During several iterations, the chromosomes evolve and new generations (offsprings) are generated as a result of crossover and mutation techniques. Crossover operator splits the two chromosomes and then combination of one half of each chromosome takes place with each pair. Mutation process involves flipping of single bit of a chromosome. The generated chromosomes are evaluated using fitness criteria and as such the best chromosomes are maintained and the others are discarded. The process is repeated until one chromosome has the fitness and is therefore considered as the best possible solution for the problem.

The objective functions in the present study are $f(x_1)$ and $f(x_2)$ and have been represented by Equations (8.1) and (8.2) for the MRR and the SR. As genetic algorithm minimizes all the solutions, in order to maximize $f(x_1)$ and minimize $f(x_2)$, Equations (8.1) and (8.2) have been modified to yield the following equations:

$$\text{Objective function} 1 = -(Y_{MRR})$$

$$\text{Objective function} 2 = (Y_{SR})$$

Size of the population, rate of mutation, number of generations i.e., iterations, etc. are the critical parameters for the genetic algorithm process. For the case under consideration, a population size of 100, rate of crossover as 90%, mutation rate of 10%, mutation function of uniform, number of iterations as 500 and population type of double vector have been considered.

TABLE 8.2
Optimal Solutions for MRR and SR

Exp. No.	Process Parameters with Un-Coded Value				Responses	
	Feed Rate (mm/min)	Power Rating (W)	Slurry Concentration (g/L)	Grit Size (µm)	Ra (µm)	MRR (g/min)
1	0.86	486	40	55	0.56	0.2360
2	0.83	486	40	57	0.67	0.2433
3	1.18	486	41	17	−0.31	0.1665
4	1.10	486	41	35	−0.07	0.1918
5	1.08	486	40	38	0.03	0.1918
6	1.11	486	41	43	0.12	0.2059
7	0.93	486	41	39	0.07	0.2007
8	1.15	486	41	34	−0.06	0.1889
9	1.09	486	41	23	−0.22	0.1739
10	0.87	486	41	50	0.40	0.2251
11	0.86	486	40	46	0.26	0.2170
12	0.94	486	40	40	0.05	0.2025
13	0.86	486	40	52	0.44	0.2292
14	0.83	486	40	56	0.68	0.2400
15	1.15	486	41	33	−0.12	0.1871
16	0.90	486	40	57	0.65	0.2410
17	0.96	486	40	52	0.43	0.17
18	1.01	486	40	41	0.08	0.15
19	1.05	486	40	37	0.00	0.14
20	1.06	486	40	39	0.02	0.14
21	0.95	486	40	29	−0.17	0.13
22	0.98	486	40	45	0.19	0.16
23	1.03	486	40	54	0.52	0.18
24	0.97	486	40	56	0.60	0.18
25	1.02	486	40	48	0.30	0.16

The problem when optimized using the genetic algorithm, the SR as well as the MRR are minimized and maximized, respectively. The combination of parameters leading to the values have been depicted in Table 8.2. As it can be observed from the table that the best possible combination of parameters for achieving optimized MRR and SR are: feed rate (mm/min): 0.86, power rating (W): 486, slurry concentration (g/L): 40 and grit size (µm): 55.

8.6 CONCLUSION

Present chapter highlights the application of genetic algorithm to address an engineering optimization problem. The problem under consideration pertains to the ultrasonic machining of hole surface. Thus the importance of genetic algorithm in addressing such problems has been demonstrated.

9 Engineering Problem Optimized Using Particle Swarm Optimization Algorithm

9.1 INTRODUCTION

Electrical discharge machining (EDM) is a non-traditional machining process that is based on the electro-thermal mechanism for material removal. Material removal takes place by means of repeated electrical discharges taking place between the electrode and the workpiece material being machined in the presence of dielectric fluid. EDM has been employed widely in the machining industry for machining of conductive materials such as graphite, metallic alloys, metals and even ceramics.

Following machining characteristics aid in evaluating the performance of the EDM process: electrode wear ratio and material removal rate. However, proper selection of machining parameters is required to achieve higher material removal rate and lower electrode wear ratio. Various techniques are employed, apart from the past experiences, to achieve optimum response parameters for machining.

A number of optimization tools have been used from time to time to address the aforementioned issue associated with the EDM process. As for instance gray-fuzzy-based Taguchi technique (Lin et al., 2002), artificial neural network (Su et al., 2004), non-dominated sorting genetic algorithm II (NSGA II) (Mandal et al., 2007), genetic algorithm (GA) (Rao et al., 2009), desirability-based simulated annealing (Yang et al., 2009), backpropagation neural network (BPNN) (Al-Anzi and Allahverdi, 2007) and particle swarm optimization (PSO) (Baskar et al., 2005). A number of PSO variants have been employed and comparative analysis has been made such as particle swarm optimization-original (PSO-O), particle swarm optimization-constriction factor (PSO-CF) and particle swarm optimization-inertia weight (PSO-IW).

As far as the context of present chapter is concerned, applicability of PSO algorithm has been demonstrated for EDM process. PSO has been delineated to optimize the machining of complex shapes in stainless steel workpiece material.

9.2 EDM PROCESS DETAILS

There are large number of factors that affects the machining performance of the EDM process. However, there are three most important process factors that critically impacts the heat input, material removal rate and electrode wear rate. These parameters are pulse-on duration (T ON), supply current (I) and pulse-off duration

TABLE 9.1

Important Process Parameters and Their Levels

	Levels		
Factors	−1	0	1
$I(A)(X_1)$	2	6	10
T (ON) (X_2)	7,000	9,000	11,000
T (OFF) (X_3)	7,000	9,000	11,000

(T OFF). Therefore these three factors are considered as design factors. A large number of pilot experiments are then run to arrive possible working ranges of the process parameters. During the pilot experiments each of the factor was varied while the rest were kept constant. The working range for each of the parameter was then decided by taking into consideration the lowest limit of power for which the material removal process can take place, and on the other hand by considering the maximum power that electrode can bear so that the extreme wear of the tool can be avoided. As such the lower power factor is coded as (−1) and the maximum power as (+1). The levels of critical factors were determined as a result of pilot experiment. These have been tabulated in Table 9.1.

9.3   EXPERIMENTAL DETAILS

The experimental was carried out using die-sinking EDM machine. EDM oil was employed as dielectric fluid. Box–Behnken design was used for designing the experiments. Holes were machined on the workpiece material. Copper was used as the electrode. The values of material removal rate and electrode wear rate were obtained by using the shop-floor data as obtained during the experimentation process. Following equation as such were employed (Ramasamy et al., 2002):

$$VMRR = \frac{MWBM - MWAM}{\rho_w \times t} \tag{9.1}$$

$$VMRE = \frac{MEBM - MEAM}{\rho_E \times t} \tag{9.2}$$

$$EWR = \frac{VMRE}{MRR} \tag{9.3}$$

where, VMRR represents the volume of workpiece material removed per unit time, VMRE is the volume of material removed from electrode per unit time, MWBM and MWAM are the mass of material removed from workpiece material before machining and that after machining, respectively, MEBM and MEAM are the material removed from electrode before and after machining, ρ_w and ρ_e are the density of the

TABLE 9.2
Matrix for Box–Behnken Response Surface Design and Measured Output Responses

S. No.	Pulse-on Time T (ON) (µs)	Supply Current I (A)	Pulse-Off Time T (OFF) (µs)	Volumetric EWR (%)	MRR (m³/sec)
1	9,000	6	9,000	5.884	8.11
2	11,000	6	7,000	4.381	2.2163
3	9,000	6	9,000	6.12	8.327
4	7,000	2	9,000	4.6815	3.318
5	9,000	10	7,000	14.376	4.119
6	9,000	2	11,000	5,387	3.514
7	7,000	6	7,000	4.24	5.114
8	11,000	10	9,000	6.442	4.808
9	11,000	6	11,000	4.6392	4.71
10	9,000	2	7,000	4.9890	3.14
11	9,000	6	9,000	5.87	8.31
12	11,000	2	9,000	4.166	3.16
13	7,000	10	9,000	16.41	4.632
14	9,000	10	11,000	4.2860	10.100
15	7,000	6	11,000	6.616	4.3218

workpiece material and the electrode, respectively, t is the actual time of machining and EWR is the electrode wear ratio.

The values for material removal as well the electrode wear ratio have been tabulated in Table 9.2.

9.4 RESPONSE SURFACE METHOD FOR EMPIRICAL MODELS

Method of response surface methodology involves the combination of statistical techniques as well as mathematical equations that aid in analysis of problems has several independent variables affecting the dependent variable (Puertas et al., 2004). However, for the practical application of RSM, it becomes necessary to develop an approximate model for the true response surface. The approximate modeling is achieved on the basis of data obtained from the system or the process and is an empirical model. Multiple regression analysis is a statistical technique that is useful in constructing such empirical models that are required for response surface methodology. Following second-order polynomial equation is generally used in response surface methodology:

$$Y = a_0 + \sum_{j=1}^{k} a_j x_j + \sum_{j=1}^{k} a_{jj} x_j^2 + \sum_{i<j} \sum_{j=2}^{k} a_{ij} x_i x_j \tag{9.4}$$

where, the regression coefficients are represented by $a_0, a_j, a_{jj}, a_{ij} = 0, 1, \ldots, k$.

Different statistical software packages are employed to measure the responses and hence determine the empirical models with best fit. The empirical models for material removal rate and electrode wear ratio are represented using Equations (9.5) and (9.6), respectively:

$$MRR = 8.32959 + 1.43977X_1 + 0.00497X_2 + 1.05594X_3 - 2.64629X_1^2 - 2.99379X_2^2$$

$$- 2.09380X_3^2 + 0.07275X_1X_2 + 1.86100X_2X_3 + 0.55764X_3X_2 \qquad (9.5)$$

$$ERR = 5.8830 + 2.6503X_1 + 1.4603X_2 + 0.7715X_3 + 2.0743X_1^2 - 0.2136X_2^2$$

$$- 0.7088X_3^2 - 2.2017X_1X_2 - 2.6210X_2X_3 - 0.5301X_3X_2 \qquad (9.6)$$

9.5 ACCURACY CHECK FOR THE MODEL

Analysis of variance (ANOVA) technique was used for testing the developed model. The calculated F-ratios were revealed to be greater than the values tabulated in Table 9.1. The confidence level was 95% and as such the developed models are considered to be adequate (Majumder, 2013).

Coefficient of determination is another criterion that is used commonly to demonstrate the accuracy of the fitted regression model. The goodness of fit is determined through the calculated values of determination coefficient as well as the adjusted determination coefficient. The values obtained were, respectively, 80% and 70% indicating a high significance of the developed model.

The accuracy of the models developed was tested through the calculated p-values. A value greater than 0.5 indicates a high significance for the developed models, while the model is highly insignificant if the value is greater than 0.1. For the case under consideration the values were less than 0.5 and as such the models were highly significant and acceptable.

9.6 OPTIMIZATION WITH PSO

PSO is a computational stochastic global search optimization method that is based on the movement of swarms to solve the optimization problem within the constraints. Kennedy and Eberhart (Al-Anzi and Allahverdi, 2007) were the ones to propose the PSO method in 1995. There are multiple candidate solutions that coexist and concurrently collaborate. Each of the solutions in the form of particle flies within the search space and continuously look for the optimal solution. The particles readjust their position in accordance with their experience as well as that of the experience of the neighboring articles. Experience of the particle is built through the tracking and memorizing process in each iterations. In PSO, there is a combination of both the global and local search experience. As such the PSO tries to balance the exploitation and exploration phases.

Let there be a search space having dimension d. The ith particle of swarm can be represented through d dimensional vector $x_i = (x_{i1}, x_{i2, ...,} x_{id})$. The velocity of the particle is represented by $v_i = (v_{i1}, v_{i2}, ..., v_{id})$. Let the best visited position for the particle be represented by p_{id} and the best position explored so far is represented by g_{id}. The velocity update rules are discussed herewith:

The initial stage of development involves the change in speed and position of each of the particle in PSO. This is done in accordance with the following equations:

$$v_{id}^{i+1} = v_{id}^i + C_1 r_1 \left(p_{id} - x_{id}^i \right) + C_2 r_2 \left(g_{id} - g_{id}^i \right) \tag{9.7}$$

$$x_{id}^{i+1} = x_{id}^i + v_{id}^{i+1} \tag{9.8}$$

where, $C_1 = 2$ is the cognitive parameter, $C_2 = 2$ is the social parameter, r_1 and r_2 are the random numbers distributed uniformly in the range [0–1] and $j = 1, 2, ...$ are the iterations.

In case of the PSO method with inertia weight, an inertia weight 'w' is associated with the last iteration velocity and as such the equations for speed and position are as under:

$$v_{id}^{i+1} = w v_{id}^i + C_1 r_1 \left(p_{id} - x_{id}^i \right) + C_2 r_2 \left(g_{id} - g_{id}^i \right) \tag{9.9}$$

$$x_{id}^{i+1} = x_{id}^i + v_{id}^{i+1} \tag{9.10}$$

As such the current velocity is influenced by the weight associated with the old velocity. Searching ability of PSO as a whole increases with the inertia weight w, as such smaller value of inertia weight signifies larger PSO ability for the partial. Generally inertia weight w ranges 0.9–0.4. This makes the PSO method to search at the beginning and locate the position quickly in the space wherein the optimist solution lies (Bai, 2010). The inertia weight could be arrived at by using the following equation:

$$w = w_{max} - \frac{w_{max} - w_{min}}{N_{max}} \times iter \tag{9.11}$$

where, w_{max} is the initial weight and is taken to be equivalent to 0.9, w_{min} is the final weight and is taken to be equal to 0.4, N_{max} is the maximum number of iterations.

In another PSO case, wherein the constriction agent has been introduced (Clerc et al., 1999). As such the formula for the position and the speed can be written using the following equation:

$$v_{id}^{i+1} = k \left\{ v_{id}^i + C_1 r_1 \left(p_{id} - x_{id}^i \right) + C_2 r_2 \left(g_{id} - g_{id}^i \right) \right\} \tag{9.12}$$

$$x_{id}^{i+1} = x_{id}^i + v_{id}^{i+1} \tag{9.13}$$

where, constriction factor $k = \dfrac{2}{\left| 2 - C - \sqrt{C^2 - 4C} \right|}$, and $C = C_1 + C_2$.

The number of iterations taken by the PSO method with constriction factor was lesser than that of the standalone PSO method in reaching to the optimum solution.

Also lesser number of iterations to convergence was taken by PSO with constriction factor in comparison to the PSO method with inertia weight. Furthermore, it was revealed that the original PSO doesn't has the ability to converge toward the optimum solution. The reason may be attributed to the absence of inertia weight that aids in controlling the previous and hence the current speed. As such the search process for the particles will slow down. Also the predictive capability of PSO with constriction factor and inertia weight was better than the original PSO.

The optimal setting of parameters is achieved that can minimize EWR and maximize MRR for the EDM process under consideration.

Therefore the desirability function was transformed to individual desirability index of the corresponding responses. Following steps are taken into consideration for arriving at the desirability index:

First the desirability index $(\hat{y}_i)$ is obtained for each of the responses. Following equations are used if maximization of the response is required:

$$\left.\begin{aligned} \hat{y}_i &= 0 && \text{if } i < S_i \\ _i\hat{y}_i &= \left[(i - S_i)/(H_i - S_i)\right]^{r_i} && \text{if } S_i \leq i \leq H_i \\ \hat{y}_i &= 1 && \text{if } i > H_i \end{aligned}\right\} \tag{9.14}$$

If the minimization of response is sought, then the individual desirability index can be obtained using the following:

$$\left.\begin{aligned} \hat{y}_i &= 0 && \text{if } i > H_i \\ _i\hat{y}_i &= \left[(i - S_i)/(H_i - S_i)\right]^{r_i} && \text{if } S_i \leq i \leq L_i \\ \hat{y}_i &= 1 && \text{if } i < S_i \end{aligned}\right\} \tag{9.15}$$

If a particular target T is required to be achieved by response, then the individual desirability index can be obtained using the following:

$$_i\left.\begin{aligned} \hat{y}_i &= 0 && \text{if } i < S_i \\ \hat{y}_i &= \left[(i - S_i)/(T_i - S_i)\right]^{r_i} && \text{if } S_i \leq i \leq L_i \\ \hat{y}_i &= \left[(i - S_i)/(T_i - H_i)\right]^{r_i} && \text{if } T_i \leq i \leq H_i \\ \hat{y}_i &= 0 && \text{if } i > H_i \end{aligned}\right. \tag{9.16}$$

where, predicted value of the ith response is represented by i, the weight exponent is represented by r_i, smallest and highest acceptable value for the ith response is represented by S_i and H_i, respectively.

In the next step, global desirability index is obtained by combining the individual desirability indices. Equation (9.17) is used for obtaining the global desirability index.

$$D = \left(\hat{y}_1^{w_1} \times \hat{y}_2^{w_2} \times \cdots \times \hat{y}_n^{w_n} \right)^{1/\sum_{j=1}^{n} w_j} \tag{9.17}$$

where, w_i is the weight associated with the individual jth response and n is the total number of response parameters.

Using the concept of fitness function, Y for the present study can be defined using the following set of equations:

$$\hat{y}_1 = \frac{MRR - MRR_{min}}{MRR_{max} - MRR_{min}} \tag{9.18}$$

$$\hat{y}_2 = \frac{EWR_{max} - EWR}{EWR_{max} - EWR_{min}} \tag{9.19}$$

$$DF = \left(Y_1^{w_1} \times Y_2^{w_2} \right)^{\frac{1}{(w_1 + w_2)}} \tag{9.20}$$

$$Y = \frac{1}{1 + DF} \tag{9.21}$$

where, the weights importance for MRR and EWR are represented by w_1 and w_2, respectively. The minimum and maximum values of MRR are represented by MRR_{min} and MRR_{max}, respectively. Similarly the minimum and maximum values of EWR are represented by EWR_{min} and EWR_{max}, respectively. Since both the MRR and EWR are important equally, $w_1 = w_2 = 0.5$. Desirability function is represented by DF. The main objective is therefore to choose an optimal parametric values that can result in maximizing the desirability function DF or minimize Y.

Table 9.3 depicts the comparison analysis between the PSO algorithm with constriction factor and inertia weight.

Validation is done by means of experimentations wherein the experiments are carried out using optimal process parameters. The values of MRR and EWR were obtained from the experiments and the percentage error between the predicted values and the results from experimentation was obtained. The calculated error

TABLE 9.3

Comparative Analysis for the PSO with Constriction Factor and PSO with Weighting

		Initial Condition	Optimal Condition
Process parameters	Pulse current, I (A)	6	6
	Time-on duration, T (ON) (unit)	9,000	9,000
	Time-off duration, T (OFF) (unit)	9,000	11,000
Response parameters	MRR (mm³/min)	8.02	8.1
	EWR	5.883	4.1804
Desirability index		0.4465	0.4355

value was revealed to be very small which confirmed the validity of the results obtained. The values for percentage error were found to be 2.18% and 6.84%, respectively.

9.7 CONCLUSION

The present chapter delineates the applicability of PSO method to the EDM non-traditional machining process. The optimal combination of parameter values was reported that resulted in better performance of the EDM process. The response factors considered were MRR and ERR. RSM was employed to establish empirical model between the process parameters and the output responses. It was revealed through the experimentation that PSO had lesser convergence capability in comparison to the PSO variants with constriction factor and inertia weights. Furthermore, the performance prediction was more accurate for the PSO with constriction factor in comparison to the original PSO and PSO with inertia weights.

10 Engineering Problem Optimized Using Ant Colony Optimization Algorithm

10.1 INTRODUCTION

Presence of higher tool wear, cutting force and poor quality of surface are some of the observable effects during machining of materials with higher strength, lower thermal conductivity and higher ductility. These adverse effects can, however, be eliminated through the usage of cutting fluids. However, there are a wide range of cutting fluids that are available in the market. However, usage of such oils has been a great concern to the environmental as well as health impacts. Therefore efforts have been made by research community to minimize the environmental impacts through various arrangements (Marksberry, 2007; Sanchez et al., 2010; Fratila and Caizar, 2011; Zhang et al., 2012). Another possible way to reduce the environmental hazards is to avoid the specific contents causing these hazards. This has been achieved by replacing the petroleum-based cutting fluids with the bio-based ones. Bio-based cutting fluids have been reported to have higher biodegradability and lubrication properties (Ozcelik et al., 2011a, 2011b).

Studies have revealed a wider impact of the vegetable-based cutting fluids on numerous mechanical processes. A number of important factors such as tool wear, tool life, torque, quality of surface have been taken into consideration while judging the impact of vegetable-based cutting fluid (Rahim and Sasahara, 2011; Kuram et al., 2013; Cetin et al., 2011).

Computer numerically controlled tools have gained popularity owing to the realization of full automation in machining. Lesser operator input, enhanced productivity and increased quality of machine part are some of the advantages of computer numerically controlled machine tools. Milling has been considered as one of the most efficient machining processes that is employed mainly when the material removal operation is being encountered. It has been widely used to mate with other part in aerospace, machinery design, automotive as well as manufacturing units. Owing to the importance of milling operation, selection of reasonable parameters for milling becomes critical, and the machining economics, safety and quality can be satisfied.

Manufacturing industries have been confronted with determination of efficient machining parameters and are still the subject of research for the manufacturing personnel. Optimum machining parameters play a critical role in not only defining the process competitiveness in the market but also aids in maintaining environmental balance. Furthermore, optimum parameter setting also aids in machining of a new product that is required to be introduced in the market.

However, the process of optimization becomes complicated as more and more parameters are being involved and more than one objectives need to be optimized. As for instance the maximization of rate of production with minimized product cost. Such a problem is known as multi-objective optimization problem. In the present chapter, optimization of milling process has been achieved. The main objective is to achieve process with minimum surface roughness, minimum specific energy and maximized tool life. The optimization has been achieved using ant colony optimization (ACO) method.

10.2 EXPERIMENTATION OF THE MILLING PROCESS

D-optimal method has been employed to design the experiments. D-optimal method is a response surface methodology that is used for conducting the experiments and hence the empirical modeling. The process has some advantages over the response surface methodology. A smaller number of experimental runs are provided by the response surface method and categorical factors can be tackled by the method.

Optimal designs have been preferred among the various available experimental design alternatives meant for second-order models. This is because of their successful applicability to mixed type mixed level factorial experimentation. D-optimal method has been considered to be one of the most popular method that seeks to maximize the determinant associated with information matrix. Furthermore, the method also provides the best estimation for the model parameters. The statistical results for the D-optimal plan obtained have been delineated in Table 10.1.

There are three numerical factors and a categorical factors for the optimization problem under consideration. Feed rate, depth of cut and cutting speed are the numerical factors whereas cutting fluid is qualitative factor. A second-order optimization model has been chosen to model and optimize the objective function. Following equation provides the form for full second-order model for the case comprising of three numerical factors and one categorical factor:

TABLE 10.1
Statistical Results of the D-Optimal Plan

Statistics	Value
Trace of $(X'X)^{-1}$	2.469
Determinant of $(X'X)^{-1}$	1.084E-18
Condition number of coefficient matrix	7.31
Maximum prediction variance (at a design point)	0.810
Average prediction variance	0.556
Scaled D-optimality criterion	2.277

$$Y = \beta_{0,k} + \sum_{i=1}^{3} \beta_{i,k} X_i + \sum_{i=1}^{3} \sum_{i=1}^{3} \beta_{i,k} X_i X_j + \sum_{i=1}^{3} \beta_{i,k} X_i^2 \qquad (10.1)$$

where, k is the cutting fluid index, numeric factors represented by i and j.

As can be observed from Equation (10.1), the model consists of quantitative as well as qualitative factors, second-order terms and two-term interactions.

A total of 17 parameters are there for the second-order model. A total of 27 experiments have been chosen to allocate 4 and 5 degrees of freedom for pure error and lack of fit. A number of algorithms have been proposed to identify the optimum design. For the case under consideration, point exchange algorithm has been employed. A random design is created in case of point exchange algorithm from the possible candidate set. Creation of new designs takes place iteratively wherein the individual points are replaced with the best available replacement and the process is repeated until no improvements are possible. The entire process is repeated for several iterations until and unless no improvements are possible. After the process is terminated, the best overall design from the inception of the process is selected (Myers et al., 2009). The candidate points used by the program consist of 18 interior points, 12 center of edges, 8 vertices and 1 overall for each level of the categorical factor. For the case study under consideration, Table 10.2 depicts the low and high levels of factors that have been selected on past and practical considerations. The D-experimental plan is depicted in Table 10.3. This is shown in terms of actual factor values 1, 2 and 3.

The machining process consists of the supply of cutting fluids through two nozzles. The flow rate of coolant for each nozzle has been held constant at 8 L/min. The step over is the selected as 12 mm and 1,050 mm is the length of the cutting path. Down milling method has been utilized because of some pros such as lesser tool wear, lesser generation of heat and better quality of surface finish.

The tests were conducted using CNC machine with steel as workpiece material. Two inserts were utilized to conduct the milling experiments. The inserts were coupled with shrink holder. The inserts used are changed periodically after the completion of milling operation. Measurement of flank wear is done for the inserts and the average is taken for the values. Measurement of flank wear is done through optical microscope. The value is measured by interrupting the experiment. Rate of tool wear is obtained by dividing wear by milling time. The tool life is obtained using Equation (10.2).

$$T = \frac{0.3}{V_{VB}} \qquad (10.2)$$

TABLE 10.2
Numeric Levels of Factor

Factors	Low Level	High Level
Cutting speed (m/min)	150	200
Feed rate (mm/rev)	0.30	0.40
Depth of cut (mm)	0.25	0.50

TABLE 10.3
D-Optimal Experimental Design

Cutting Speed (m/min)	Depth of Cut (mm)	Feed Rate (mm/rev)	Case Type
200	0.25	0.37	1
150	0.32	0.40	1
200	0.45	0.40	1
150	0.25	0.30	1
200	0.35	0.30	1
150	0.45	0.35	1
166.67	0.38	0.30	1
150	0.32	0.40	1
150	0.25	0.30	2
150	0.45	0.40	2
166.67	0.45	0.40	2
200	0.45	0.30	2
200	0.25	0.40	2
175	0.25	0.30	2
150	0.45	0.30	2
187.50	0.35	0.35	2
150	0.35	0.35	2
166.67	0.45	0.40	3
166.67	0.25	0.40	3
200	0.38	0.40	3
150	0.45	0.40	3
200	0.25	0.30	3
183.33	0.45	0.30	3
150	0.35	0.30	3
166.67	0.25	0.33	3
183.33	0.45	0.30	3
150	0.45	0.40	3

where, V_{VB} is the rate of tool wear in mm/min and T is the tool life in min. Maximum flank wear is 0.3 mm in finish milling process.

Dynamometer is used to measure the generated force which is mounted under the workpiece. The resultant cutting force is computed using Equation (10.3).

$$F_C = \sqrt{F_x^2 + F_y^2 + F_z^2} \tag{10.3}$$

Equation (10.4) is then used to arrive at the cutting power:

$$P_c = F_c \cdot v \tag{10.4}$$

where, P_c is the cutting power utilized for milling in W, v is the cutting speed in m/s and F_c is the cutting force in N.

The specific energy, on the other hand, is obtained using Equation (10.5):

$$U = \frac{P_c}{MRR} \qquad (10.5)$$

where, U is the specific energy in J/mm^3 and MRR is the material removal rate in mm^3/s.

Energy consumed during the machining process determines the environmental impact that the machining process has. Specific energy is often used to quantify the energy consumed during the material removal process and as such specific energy has been considered in this study.

10.3 OPTIMIZATION

Ant colony optimization is the technique of designing algorithms for optimization problems. The method is based on the cooperative behavior of the real colonies of ant that have the characteristic feature to find the shortest possible path from their nest to the food source. Continuous ant colony optimization is one of the first technique to aid in optimization of the continuous optimization problem. The steps of chaotic ACO (CACO) algorithm have been described as follows:

10.3.1 SET THE INITIAL VALUES

Assume that there is a suitable number of ants in the colony. A random vector for a continuous problem with n variables is assigned to each ant and is represented by the vector $X(x_1, x_2, ..., x_n)$. A pheromone weight τ_0 is attached to every variable in the ant. After the evaluation is made on each ant and sorting is done from the best to the worst, the pheromone on variable j of ant I is modified in accordance with the ant's rank k_i as follows:

$$\tau_{ij} = \tau_{ij} + \alpha(GN + 1 - k_i)\tau_0 \qquad (10.6)$$

10.3.2 SELECTION

A set of values on each of the variable is obtained after the initialization process. The ant is copied into the next iteration if the random number p_1 is larger than first selection probability q_1. Next, the reconstruction of ant takes place by selecting the values for each variable from the corresponding set. A pseudorandom rule of the ant colony system is followed in the reconstruction process. The rule followed includes the exploration and exploitation mechanism and the selection of which is determined by the second selection probability q_2.

The probability of the value being picked by the roulette wheel selection can be defined using the following equation:

$$p(x_{ij}) = \frac{\tau_{ij}}{\sum\limits_{i=1}^{M} \tau_{ij}} \qquad (10.7)$$

10.3.3 DUMPING OPERATION AND PHEROMONE UPDATE MECHANISM

During the dump operation, new materials for the algorithm are imported. The worst ants are replaced by the new ants with solutions being produced randomly. The algorithm then updates the pheromones on each solution in accordance to its new rank k_i. This is done in accordance with the following Equation (10.8):

$$\tau_{ij} = \rho\tau_{ij} + \alpha(GN + 1 - k_i)\tau_o \qquad k_i < GN$$
$$\tau_{ij} = \rho\tau_{ij} \qquad\qquad\qquad\qquad \text{otherwise} \tag{10.8}$$

Pheromone on the best Generalized Net (GN) solution are emphasized after evaporation owing to their reduction through multiplication with evaporation rate ρ.

10.3.4 RANDOM SEARCH

A n dimension radius R represented as $(r_1, r_2, ..., r_n)$ is related with the dimension vector X represented by $(x_1, x_2, ..., x_n)$ during the process of random search x_j has equivalent chances of hitting the values of $(x_j - r_j, x_j)$, x_j and $(x_j, x_j + r_j)$. The random search process then generates new solutions and the number of such generated solutions is decided by a predefined constant. A comparative analysis between the original solution and the solution generated by random search process is made by the algorithm. The radius will be increased to r/ϕ if the best generated solution is better than the original solution. Here, ϕ is the changing rate in the interval $(0,1)$. On the other hand, the radius r will be reduced to ϕ in case the original solution is better and hence the search space will shrink. The precision of the algorithm is ensured with this adaptive searching radius. Random search is performed only on the best ant in order to save the computational cost.

Therefore, from the above discussion, it is clear that the CACO algorithm iterates itself within three steps: selection, dump operation and pheromone update, and random search. The steps are repeated until it reaches the termination condition. Then the result is the solution of the best ant in the history.

For the case under consideration, response models have been considered to optimize the end milling operation. The models obtained are depicted using Equations (9–11).

$$U = \psi_{o,k} + \sum_{i=1}^{3}\psi_{i,k}X_i - 5.12Vf + 1.15Va_p - 1660.38f^2$$

$$+ 284.23a_p^2 + 9.70Vf^2 - 1.51Va_p^2 \tag{10.9}$$

$$R_a = \psi_{o,k} + \sum_{i=1}^{3}\psi_{i,k}X_i - 0.006Vf + 0.010Va_p - 0.38fa_p$$

$$+ 7.46 \times 10^{-5}V^2 - 5.31f^2 + 3.52a_p^2 \tag{10.10}$$

$$T = \psi_{o,k} + \sum_{i=1}^{3} \psi_{i,k} X_i + 0.49Vf + 261.67f^2 \qquad (10.11)$$

The coefficients of the linear terms in the respective models have been depicted in Table 10.4.

The responses may be utilized alone in the optimization problem. The case under study qualifies for constrained optimization with both categorical and numerical values. The general model has been depicted using Equation (10.12):

$$\text{Optimize } y = f(X, Z)$$
$$\text{Subjected to } L_i \le X_i \le U_i, ii = 1, 2, 3 \qquad (10.12)$$

Generation of energy has been considered to be a main contributor to the carbon dioxide emission and as such climate change. Thus reduction of energy consumption is one of the essential considerations in sustainable manufacturing (Hanafi et al., 2012). Hence one of the main objectives of the present study is the minimization of the specific energy consumed. This has been solved using the aforementioned discussed CACO method. The results obtained have been summarized in Table 10.5.

TABLE 10.4

Coefficients of Linear Terms in Respective Response Models

Response	Case	$\psi_{0,k}$	$\psi_{1,k}$	$\psi_{2,k}$	$\psi_{3,k}$
Specific energy	I	−61.03	0.373	680.52	−172.63
	II	−59.52	0.365	680.52	−172.63
	III	−57.07	0.356	680.52	−172.63
Surface roughness	I	2.60	−0.019	2.75	−2.38
	II	2.56	−0.018	2.30	−2.81
	III	2.46	−0.017	2.68	−2.88
Tool life	I	42.33	−0.15	−218.66	−17.11
	II	45.49	−0.15	−221.02	−17.11
	III	47.65	−0.15	−233.03	−17.11

TABLE 10.5

Results for Optimization of Specific Energy

Case Type	Cutting Speed (m/min)	Feed Rate (mm/rev)	Depth of Cut (mm)	Specific Energy (J/mm³)	Surface Roughness (µm)	Tool Life (min)
1	150	0.30	0.35	1.4311	0.4142	3.5881
2	150	0.30	0.35	2.8211	0.3998	4.8989
3	175	0.35	0.35	1.8496	0.4926	1.8138

TABLE 10.6
Results for Optimization of Surface Roughness

Case Type	Cutting Speed (m/min)	Feed Rate (mm/rev)	Depth of Cut (mm)	Specific Energy (J/mm³)	Surface Roughness (μm)	Tool Life (min)
1	175	0.30	0.258	2.2144	0.3124	3.6938
2	170	0.30	0.342	3.3831	0.3741	3.8649
3	166	0.30	0.350	2.5888	0.3946	4.3494

TABLE 10.7
Results for Optimization of Tool Life

Case Type	Cutting Speed (m/min)	Feed Rate (mm/rev)	Depth of Cut (mm)	Specific Energy (J/mm³)	Surface Roughness (μm)	Tool Life (min)
1	150	0.30	0.25	1.7411	0.3964	5.3878
2	150	0.30	0.25	3.3530	0.4397	6.6986
3	150	0.30	0.25	2.8827	0.4698	6.9791

Another objective of the study is the minimization of surface roughness and the results obtained have been tabulated in Table 10.6. Table 10.7 depicts the minimization results obtained for the maximization of tool life.

The optimization for the milling process can be considered in a more general form i.e., different objectives may be defined by considering the performance of various processes and hence multi-objective optimization can be considered. As for instance, the surface roughness determines the surface quality of the finished product. Furthermore, the other processes that determines the economy of the end milling are tool life, tool wear and other tool-related responses. The specific energy, on the other hand, reflects the cost of electrical energy. The aforementioned cost items as such forms the part of direct operating cost.

There are numerous ways in which a multi-objective optimization problem may be defined. In the present chapter, the problem has been defined in context of implicitly constrained optimization problem. One of the objective function y is considered to be optimized while the others are used as implicit constraints:

$$\text{Optimize } y = f(X, Z)$$
$$\text{Subject to}: L_i \leq X_i \leq U_i, ii = 1, 2, 3$$
$$L_i \leq f_i \leq U_i, ii = 1, 2, 3$$

where, L and H are the lower and upper bounds on the implicit constraints. These are selected on the basis of mono-objective optimization results. As for instance:

$U \leq 2.5 \text{ J/mm}^3$
$R_a \leq 0.2 \, \mu\text{m}$
$T \geq 3 \, \text{min}$

TABLE 10.8
Results for Multi-Objective Optimization When Specific Energy is Primary Objective Function

Case Type	Cutting Speed (m/min)	Feed Rate (mm/rev)	Depth of Cut (mm)	Specific Energy (J/mm³)	Surface Roughness (µm)	Tool Life (min)
1	150	0.30	0.318	1.5168	0.3998	4.0639
2	150	0.30	0.350	2.8211	0.3998	4.8989
3	170	0.30	0.318	2.7094	0.3999	4.6063

TABLE 10.9
Results for Multi-Objective Optimization When Surface Roughness is Primary Objective Function

Case Type	Cutting Speed (m/min)	Feed Rate (mm/rev)	Depth of Cut (mm)	Specific Energy (J/mm³)	Surface Roughness (µm)	Tool Life (min)
1	170	0.30	0.262	1.8786	0.1065	4.1156
2	160	0.30	0.350	2.8734	0.1588	4.1187
3	165	0.30	0.350	2.3878	0.1824	4.1282

TABLE 10.10
Results for Multi-Objective Optimization When Tool Life is Primary Objective Function

Case Type	Cutting Speed (m/min)	Feed Rate (mm/rev)	Depth of Cut (mm)	Specific Energy (J/mm³)	Surface Roughness (µm)	Tool Life (min)
1	150	0.30	0.25	1.7410	0.1742	5.1656
2	151	0.30	0.29	2.8705	0.1873	5.6325
3	162	0.30	0.32	2.3617	0.1887	4.8801

The results obtained have been depicted in Table 10.8 for the situation when primary objective function is selected to be specific energy. Similarly, Tables 10.9 and 10.10 depict the results for the situation when surface roughness and tool life have been considered as primary objective functions, respectively.

10.4   CONCLUSION

In the present chapter, optimization of the milling process is done through CACO. The optimization has been done for surface roughness, specific energy and tool life. The D-optimal methodology was adopted to model the various responses. Cutting speed, feed rate and depth of cut were the input factors. CACO was also demonstrated to address multi-objective optimization of the end milling process by considering each of the responses as primary objective function.

11 Engineering Problem Optimized Using Bees Algorithm

11.1 INTRODUCTION

There has been growing interests among the researchers to develop laser beam machining process in order to machine the materials that are difficult to be machined using traditional machining processes (Jayswal, 2005). Laser beam machining process classifies itself as thermal energy-based process and has been used to fulfill the present day requirement that demands for higher flexibility and rates of production. The laser beam machining process allows for adaptability to automation process, eliminates machining process, improvises quality of product, processes materials irrespective of their electrical conductivity and facilitates minimum heat affected zone (HAZ). The material removal process takes place through melting, vaporization and chemical degradation mechanisms wherein the chemical bonds are broken resulting in degradation of material. The laser is absorbed by the workpiece material as it is delivered from the source of higher density laser beam. As a result, the workpiece material is melted, vaporized and changes chemically such that the material can be removed by the flow of high pressure assist gas jet. The process also doesn't involve any force of cutting and eliminates the chances of tool wear. Several material processing operations can be accomplished using the laser beam machining method. This involves laser cutting, micro-turning, micro-drilling, scribing and marking (Meijer, 2004; Dubey and Yadava, 2008a).

Among various types of lasers used, Nd:YAG and CO_2 are the most established lasers. Nd:YAG laser offers numerous advantages in comparison to the CO_2 lasers and as such makes it an interesting domain of investigation. Some unique characteristics are possessed by Nd:YAG lasers as revealed from experiments. Beam intensity is very high, although the beam power is very low. This is due to the smaller duration of pulse and better focusing behavior. Nd:YAG lasers produce parts with lower kerf width, narrower HAZ, microsize holes and better cut edge kerf profile. Owing to the smaller thermal load, the Nd:YAG lasers are capable to machine brittle materials such as ceramics and has therefore advantage over CO_2 lasers.

Numerous parameters such as lamp current, frequency of pulse, air pressure, cutting and pulse width are associated with the complex Nd:YAG laser beam machining process. Hence it is critical to select suitable combination of parameters in order to achieve higher rates of production and at acceptable level of quality. Experimentations have revealed improved performance of Nd:YAG laser beam machining process by selection of proper combination of parameters (Dubey and Yadava, 2008b). As such the process engineers are required to often rely on the handbook or manufacturer's data. Therefore, the need for ardent optimization tool cannot be ignored that allows for the suitable combination of parameters and hence the enhanced performance of Nd:YAG lasers.

Numerous studies have been reported on the parametric optimization to determine the optimal process parameters (Dhupal et al., 2007; Dubey and Yadava, 2008c; Dhupal et al., 2008; Sivarao and Ammar, 2010; Panda et al., 2011). As such the researchers have applied different optimization techniques: Taguchi method, grey relational analysis, genetic algorithm, desirability function, etc., for identification of suitable process parameters for the Nd:YAG laser beam machining process. However, these technique have yielded near optimal or sub-optimal solutions. Therefore the present chapter demonstrates the applicability of artificial bee colony optimization process.

11.2 ARTIFICIAL BEE COLONY ALGORITHM

Karaboga et al. (Karaboga and Basturk, 2007a; Karaboga and Basturk, 2008; Karaboga and Akay, 2009; Karaboga and Akay, 2009) developed artificial bee colony optimi-zation algorithm which is an evolutionary computational technique. There are three types of bees that encompasses the colony of artificial bees: employed, onlookers and scout bees. The first half of the colony is comprised of employed bees whereas the onlookers form the second half of the colony. There is only one employed bee for every source of food and the number of employed bees equals the number of food sources situated around the hives. The scout origins whenever the employed bee abandons its food source.

Possible solution in the algorithm is represented by the position of a food source and the fitness value of the function emulates the nectar quality. There is equivalent number of onlooker and employed bees. As a first step to the algorithm, random distributed predefined number of initial population P is generated. The position of food source is represented by x_{ijk} which is three-dimensional wherein $i = 1,2,...,SN, j = 1,2,...,D$ and $k = 1,2,...,V$. D is the dimension of each variable and V is the number of variables associated with the objective function. The randomly generated solutions are subjected to repeated cycles, $C = 1,2,...,MCN$ where MCN is the maximum number of cycles for the search processes of the employed, scout and onlooker bees.

Modification of the position is produced by the employed bee itself which depends on the local information as such the visual information and for this it tests the amount of nectar possessed by the new source of food. The bee will memorize the new position if it finds the nectar amount in the found new food source to be greater than the previous one. Otherwise, the bee keeps the previous position in its memory. The nectar information of the food source is shared once the entire search operation has been performed and also the position is shared to the onlooker bees in the dance arena. The nectar value is evaluated by the onlooker bees which they take from the employed bees and then selects the food source with the probability associ-ated with the amount of nectar detected. The onlooker bees produce a change in its position and memorizes this new position if the nectar amount detected is higher than that of the previous food source.

The probability with which the onlooker bees will select a food source will depend on the following:

$$p_i = \frac{\text{fit}_i}{\sum\limits_{i=1}^{\text{SN}} \text{fit}_i}$$ (11.1)

where, fit_i is the fitness value of the ith solution. The fitness value proportionate the nectar amount of the food source in the ith position and the number of food source is denoted by SN which is equivalent to the number of employed bees.

Artificial bee colony optimization produces food position from the old position by adopting the following equation:

$$v_{ijk} = x_{ijk} + \phi\left(x_{ijk} - x_{ljk}\right)$$ (11.2)

where, the candidate food solution is represented by v_{ijk}. L is determined randomly and has to be different from i, φ is the random number ranging -1 and 1. This random number controls the position of the neighborhood source around x_{ij} and also represents a visual representation of the two food sources. The perturbation of the position x_{ijk} also decreases as the difference between the two positions decreases. The step length as such is reduced adaptively as the search space approaches near optimal solution. The parameter can be set to an acceptable value if the parameter value produced by this operation exceeds the predetermined limit.

The food source wherein there is no nectar is replaced by a new food source and this is done by scout bees. Simulation in artificial bee colony algorithm (ABC) algorithm of the random position reflects the aforementioned process. This is replaced with the abandoned food site source. The food source is assumed to be abandoned in case no further improvement seems to be possible even after predetermined number of cycles. The value of number of cycles is one of the critical parameters for the ABC algorithm which is referred to as the "limit" for abandonment. A new food source is discovered by the scout as and when the previous site is abandoned. Equation (11.3) defines this operation:

$$x_{ijk} = x_{\min jk} + \text{rand}(0,1)\left(x_{\max jk} - x_{\min jk}\right)$$ (11.3)

where, $x_{\max jk}$ and $x_{\min jk}$ are the upper and lower bounds of kth variable, respectively. At each of the source position, the value of v_{ijk} is searched out and is evaluated by the bees. The performance is then compared with that of the old position. The performance of the new position is then compared with that of the old one. The new food source is replaced with the old one if the nectar amount possessed by it is in equal amount to that possessed by the new one. As such the old position is retained. The greedy selection mechanism is employed as to select between the old and the new one. The main steps of ABC algorithm are as follows (Karaboga and Basturk, 2007b):

1. Initialize
2. Repeat
   a. Place the employed bee on the food source
   b. Place the onlooker bees on the food source
   c. Send the scouts to the area of search
3. Termination as and when the criteria is met.

A pseudo code for the ABC algorithm has been presented as follows:

1. Population of the solution is x_{ijk} is initialized.
2. Evaluation of the generated population is done.
3. Follow the first cycle.
4. Repeat the cycle.
5. New solutions are generated for the employed bees. Evaluation of the solution produced is then done.
6. Greedy selection process is then applied.
7. The probability values are then calculated and evaluated.
8. New solutions for the onlookers are produced from the employed bees solution. The generated solution is evaluated for its consideration as potential future candidate solution.
9. The greedy selection process is applied again.
10. The abandoned solutions for the scouts are determined and if exists then they are replaced with the new randomly generated solutions.
11. The best solution is memorized and archived.
12. Next cycle is progressed i.e., cycle = cycle + 1.
13. Until cycle = MCN.

11.3 OPTIMIZATION OF THE ND:YAG LASER BEAM MACHINING PROCESS USING ABC

The applicability of the ABC algorithm has been demonstrated using the parametric optimization of the Nd:YAG laser beam machining process. Single as well as multi-objective optimization of the aforementioned machining process has been demonstrated using the ABC algorithm. Following values have been considered for the important control parameters of the ABC algorithm: swarm size = 10, number of onlooker bees = 50% of the size of the swarm, number of the employed bees = 50% of the size of the swarm, number of cycles considered = 2,000, number of scout bees per cycle = 1 and runtime parameter has been considered as 2.

The role of the various control parameters is critical as these parameters drive the operational aspect of the algorithm. As for instance, number of onlookers, employed bees and swarm size directly impacts the selection of the initial starting point for the algorithm. Also the number of bees participating in the search process depends upon the aforementioned parametric values. The time frame at which the evaluation of the food source is done depends upon the number of onlooker bees. This means that higher the number of onlooker bees, quicker the chances to select or reject the food sources. Furthermore, the search process for the new food will be steadfast as the number of scouts per cycle is increased. The number of cycles represents the number of times the algorithm will be required to run before the termination criteria is met. This may prove useful in situations wherein a large number of variables are required to be evaluated.

The number of iterations required to reach optimal solution increases/decreases with the number of changing values of the control parameters. However, there will be no significant changes in the optimal solution obtained. The present chapter chooses the set of parametric values on the basis of mathematical model.

The present case considers the micro-drilling of ceramics using laser beam machining process. The process has been studied for the ceramics. Influence of four important parameters: pulse width, air pressure, pulse frequency and lamp current has been studied for the taper of the hole and the thickness of the HAZ. Five different levels were chosen to study the process. These levels have been depicted in Table 11.1.

Experimentations were carried out in accordance with the central composite rotatable second-order design plan and based on response surface methodology. Experimental plan has been tabulated in Table 11.2.

Following equations as such were developed:

$$HAZ = 0.2685 + 0.06777x_1 - 0.03110x_2 - 0.04200x_3 - 0.00460x_4$$

$$+ 0.01035x_1^2 - 0.00956x_2^2 + 0.00155x_3^2 - 0.00123x_4^2$$

$$- 0.0117x_1x_2 - 0.00568x_1x_3 - 0.02247x_1x_4 - 0.00221x_2x_3$$

$$- 0.00872x_2x_4 - 0.00386x_3x_4 \tag{11.4}$$

$$TAPER = 0.06142 + 0.00801x_1 + 0.00776x_2 - 0.00505x_3 + 0.00338x_4$$

$$+ 0.00042x_1^2 + 0.00336x_2^2 + 0.00122x_3^2 + 0.00288x_4^2$$

$$+ 0.003321x_1x_2 - 0.00535x_1x_3 - 0.00418x_1x_4 - 0.00221x_2x_3$$

$$- 0.00012x_2x_4 - 0.06142x_3x_4 \tag{11.5}$$

For the case of single-objective optimization, second-order RSM-based equations have been optimized separately. In the present case the responses to be minimized

TABLE 11.1
Machining Parameters and Their Levels

Parameters	Levels				
	-2	-1	0	1	2
Lamp current (x_1) (amp)	16	18	20	22	24
Pulse frequency (x_2) (kHz)	1	2	3	4	5
Air pressure (x_3) (Kg/cm²)	0.5	0.9	1.3	1.7	2.1
Pulse width (x_4) (%)	3	7	11	15	19

TABLE 11.2

Experimental Plan for the Laser Beam Machining Process

Experiment No.	x_1	x_2	x_3	x_4	HAZ (mm)	Taper
1	20	3	1.3	11	0.4328	0.0723
2	18	2	4	7	0.2997	0.0694
3	18	2	4	15	0.2982	0.0991
4	18	2	0.9	7	0.3243	0.0605
5	22	2	1.7	15	0.4465	0.0722
6	20	3	1.3	11	0.4222	0.0991
7	22	4	0.9	15	0.3894	0.1394
8	18	4	1.7	7	0.2540	0.0743
9	22	4	1.7	15	0.3639	0.0924
10	18	4	0.9	7	0.3240	0.0858
11	18	4	0.9	15	0.3419	0.0935
12	22	2	0.9	15	0.5987	0.0988
13	18	4	1.7	15	0.3115	0.0911
14	22	2	0.9	7	0.6278	0.0992
15	22	2	1.7	7	0.5271	0.0871
16	22	4	1.7	7	0.4236	0.0999
17	18	2	0.9	15	0.4586	0.0992
18	22	4	0.9	7	0.5292	0.1168
19	20	3	1.3	11	0.3969	0.0797
20	20	3	1.3	11	0.3705	0.0752
21	20	3	1.3	3	0.3202	0.0952
22	24	3	1.3	11	0.5937	0.1116
23	20	3	0.5	11	0.4594	0.0977
24	20	5	1.3	11	0.3379	0.0992
25	20	3	1.3	11	0.3444	0.0899
26	20	3	1.3	19	0.2992	0.0996
27	20	1	1.3	11	0.3957	0.0707
28	20	3	2.1	11	0.2801	0.0749
29	16	3	1.3	11	0.2981	0.0645
30	20	3	1.3	11	0.3676	0.0644
31	20	3	1.3	11	0.3997	0.0836

should be under the following constraints: $16 \leq x_1 \leq 24$; $1 \leq x_2 \leq 5$; $0.5 \leq x_3 \leq 2.1$ and $3 \leq x_4 \leq 19$. The optimum solutions obtained for the two responses have been depicted in Table 11.3.

In the case of multi-objective optimization of Nd:YAG laser beam machining process, the two responses are optimized simultaneously. This is achieved by developing the following objective function:

$$\text{Min}(Z_1) = \frac{w_1 \text{HAZ}}{\text{HAZ}_{min}} + \frac{w_2 \text{TAPER}}{\text{TAPER}_{min}} \qquad (11.6)$$

TABLE 11.3

Results for Single-Objective Optimization

Response	Optimal Value	Lamp Current (amp)	Pulse Frequency (kHz)	Air Pressure (Kg/cm²)	Pulse Width (%)
HAZ	0.0186	16	4.7	2.0	3
Taper	0.0213	17.2	1.15	0.5	3

TABLE 11.4

Results Obtained for Multi-Objective Optimization

Response	Optimal Value	Z_1	Lamp Current (amp)	Pulse Frequency (kHz)	Air Pressure (Kg/cm²)	Pulse Width (%)
HAZ	0.1397	0.0959	16	1.72	1.05	3
Taper	0.0511					

where, HAZ and TAPER are the second-order equations based on response surface methodology for the HAZ and taper, respectively. The minimum values of HAZ and taper are represented by HAZ_{min} and $TAPER_{min}$, respectively, w_1 and w_2 are the weights associated with the HAZ thickness and the taper, respectively. The values of w_1 and w_2 can be anything such that $w_1 + w_2 = 1$. Assignment of the weights is dependent entirely on the experience and knowledge of the concerned process engineers. The weights can be determined using the analytic hierarchy process. For the present case, equal weights have been considered. The problem is solved using the ABC algorithm and the obtained results have been tabulated in Table 11.4.

11.4 CONCLUSION

In the present chapter, parametric optimization of the Nd:YAG laser beam machining process has been approached using the ABC algorithm. The results obtained for both the single as well as multi-objective optimization have been derived. Performance of ABC algorithm has been found to be better and hence the applicability of the ABC algorithm for solving non-traditional machining processes.

12 Engineering Problem Optimized Using Firefly Algorithm

12.1 INTRODUCTION

One of the most significant parts of industries associated with manufacturing is the machining process. The intricate shapes are machined using the non-traditional machining process. Engineers are required to give their best and produce products with higher performance and better effectiveness. As such umpteen efforts have been made to identify optimal solution for the machining processes using different optimization techniques such as particle swarm optimization (PSO), artificial neural network (ANN), simulated annealing (SA) and genetic algorithm. Owing to the importance of optimization techniques, researchers have developed a number of optimization techniques that depict the nature in their operation. One such algorithm is the firefly algorithm developed by Yang (2008). The algorithm makes use of the flashing light emanating from insects in terms of intensity. The firefly algorithm is based on the conceptual framework of population-based evolutionary framework and is employed to address problems associated with wide range of domains. Certain unique characteristics have been revealed to be possessed by the firefly algorithm such as higher rates of convergence, multi-modal characteristics, etc.

Optimization of machining processes has resulted in reduced cycle time and better machining. Numerous works have been reported in this regard. Owing to the subject matter of the present chapter, firefly algorithm has been employed to improve the performance parameters of electrochemical machining processes. Electrochemical machining process has been used for machining of complex geometry at micro level. The process has been employed for machining of micro holes with higher accuracy and enhanced production rate. The next section describes the firefly algorithm and subsequently its application to electrochemical machining optimization. The chapter finally concludes with the terminating remarks.

12.2 FIREFLY ALGORITHM

Fireflies have been considered as one of the wonderful creations of god because of the life style possessed by them which is quite unique in comparison to the other creatures on this planet. Yang (2008) has been credited to the development of the firefly algorithm. Fireflies dwell in warm environment and are active in summer nights. These are characterized by the flashing light produced as a result of the biochemical processes. The flashing light serves two purposes: attract mating partners and to warn the potential predatory creatures. The firefly species attract their mates through the pheromone trail. Rule of physics is followed by the flashing light i.e., its intensity decreases with the increasing distance, and the intensity is therefore in accordance with the following equation: $I = I/r^2$, where I is the intensity of the light and r is the distance. Firefly acts similar to the LC oscillator that charges and discharges at regular interval of time i.e., $\theta = 2\pi$. A mutual coupling takes place when a firefly is placed within another firefly. Most of the biochemical creatures have been revealed to provide only modulated flashes whereas firefly species are able to control biochemical reaction in order to emit high and discrete flashes. It is the signals emanating from the nervous system of firefly that produces light for the firefly species. Flying males are the first signalers who attract the females on the ground surrounding them. The females responds by continuously emitting or flashing light. The behavioral differences in the signals from the fireflies form the basis for getting attracted and as such the females will be attracted toward the firefly that flashes the brightest light. The attraction behavior is also affected by the distance between the female and male firefly because of the inverse relationship between the light intensity and the distance.

Variation in intensity as well as the attractiveness in the firefly plays a major role in firefly algorithm performance. The intensity of light decreases as the distance increases because of the absorption of the light. These phenomenon forms the major objective of the objective function to be optimized. As such relationships have been developed between different parameters that effect directly the performance of the firefly algorithm. Randomness reduction, randomness factor and absorption factors are the factors that affect the performance of the firefly algorithm. Random initialization of the population is done in the firefly algorithm by taking into consideration the bounds with initial parameters. Once the population has been initialized, the parameters at each iteration are controlled by the distance between fireflies, absorption coefficient and randomness factor. The process parameters are changed on the basis of these parameters through the evaluation of the objective function. The value of the objective function is compared at each iteration and if the value at the current iteration is found to be better than the previous one, then the solution value will be stored, otherwise the previous value will be stored rejecting the current value. All iterations are carried out in the similar manner to find out the optimal solution for the performance parameter under consideration. Maximum number of iterations control the search space and the algorithm terminates after reaching the maximum number of iteration count. Following discussion summarizes the steps involved in the firefly algorithm:

Step 1: for $t = 0$; absorption coefficient (γ) is taken as 1 and optimum value of variables $(s*)$ is equivalent to φ.

Step 2: initialization of the population at t counter i.e., $P(t) =$ Initialize Firefly algorithm population.

Step 3: Start the iteration and continue till $t < t_{max}$.

Step 4: Consider the randomization parameter i.e., $\alpha(t) =$ Alpha new ().

Step 5: Firefly intensities are evaluated next.

Step 6: Ordering of the firefly intensities are done.

Step 7: Best firefly intensity is calculated.
Step 8: Population is updated next.
Step 9: Counter is incremented by 1.
Step 10: Termination of the firefly algorithm.

The intensity of the light in case of firefly algorithm represents the fitness of the objective function under consideration. Equation (12.1) aids in determination of the light intensity:

$$I(r) = I_0 e^{-\gamma r^2} \qquad (12.1)$$

where, the intensity of the source is denoted by I_0 and the light absorption coefficient is represented by γ.

Attractiveness of the firefly, denoted by β, is proportional to the intensity of the light and Equation (12.2) is used for the determination of the attractiveness index:

$$\beta = \beta_0 e^{-\gamma r^2} \qquad (12.2)$$

where, attractiveness at $r = 0$ is denoted by β_0.

The distance between two fireflies i and j is expressed using Equation (12.3):

$$r_{ij} = \sqrt{\sum_{k=1}^{n} (s_{ik} - s_{ij})^2} \qquad (12.3)$$

where, the position vectors of the fireflies are denoted by s_i and s_j. The dimension of the problem under consideration is denoted by n, as each firefly i is attracted by firefly j, the parameters of firefly algorithm will also vary and this will be in accordance with the Equation (12.4):

$$s_i(t+1) = s_i(t) + \beta_0 e^{-\gamma r_{ij}^2} (s_j(t) - s_i(t)) + \alpha \varepsilon_i \qquad (12.4)$$

where, ε_i is a random parameter which is obtained from the Gaussian distribution. Three terms form the crux of firefly movement: current position of ith firefly and attraction to other firefly that is more attractive and randomization parameter α.

12.3 APPLICATION OF FIREFLY ALGORITHM TO ELECTROCHEMICAL MACHINING OPTIMIZATION

The present section demonstrates the applicability of firefly algorithm to solve optimization problem for the domain of non-traditional machining process. As such the firefly algorithm has been applied to optimize electrochemical micro machining process. The material removal rate of electrochemical machining process is relatively higher in comparison to the other non-traditional machining operations such as electrical discharge machining. However, the power requirement is higher. The electrochemical machining is known as electrochemical micro machining

TABLE 12.1
Process Factors and Their Limits

Parameters	Lower Bound	Upper Bound
Pulse on/off ratio (x_1)	0.5	2.5
Machining voltage (V) (x_2)	2.5	4.5
Electrolyte concentration (g/l) (x_3)	10	30
Voltage frequency (Hz) (x_4)	35	55
Tool vibration frequency (Hz) (x_5)	100	300

process when the process is applied to fabricate micro structures and to thin film domain. The cathode doesn't necessarily carry the shape of the desired contour on the workpiece surface. The 3D shaping in electrochemical micro machining process involves mask less and through mask material removal. The tool in this case of machining may be connected to computer numerical controlled machine which aids in production of complex shapes with the employability of single tool.

In the present case the machining parameters selected are: pulse on/off ratio, electrolyte concentration, frequency of tool vibration, machining voltage and concentration of electrolyte. The levels of the parameters are depicted in Table 12.1 (Munda and Bhattacharya, 2008).

Material removal rate and radial overcut are the performance parameters considered for the present work. The mathematical models for the two response parameters are depicted through Equations (12.5) and (12.6), respectively:

$$MRR = -1.79928 + 0.122969x_1 + 1.37274x_2 - 0.0975055x_3 + 0.0342233x_4$$

$$- 0.00249749x_5 - 0.302777x_1^2 - 0.0970693x_2^2 - 0.000256863x_3^2$$

$$- 0.000429643x_4^2 + 0.00000499479x_5^2 - 0.0805437x_1x_2 + 0.00949947x_1x_3$$

$$+ 0.0038677x_1x_4 + 0.00189595x_1x_5 + 0.00980375x_2x_3 - 0.00811875x_2x_4$$

$$- 0.00216115x_2x_5 + 0.00136548x_3x_4 + 0.0000358737x_3x_5$$

$$+ 0.0000292275x_4x_5 \tag{12.5}$$

$$ROC = -1.09259 + 1.32049x_1 + 0.559749x_2 - 0.0932444x_3 + 0.0358894x_4$$

$$- 0.00369699x_5 + 0.299652x_1^2 + 0.0665987x_2^2 + 0.00219558x_3^2$$

$$+ 0.000750439x_4^2 + 0.000003928306x_5^2 - 0.249977x_1x_2 - 0.00244978x_1x_3$$

$$- 0.0272869x_1x_4 - 0.000065244x_1x_5 - 0.00418692x_2x_3 - 0.0274302x_2x_4$$

$$+ 0.000942442x_2x_5 + 0.00089765x_3x_4 + 0.0000358737x_3x_5$$

$$+ 0.0000292275x_4x_5 \tag{12.6}$$

Following control parameters were adopted to solve the single-objective optimization problem using the firefly algorithm:

a. Number of iterations were considered to be 100
b. Number of fireflies were considered as 20
c. Initial randomness was taken as 0.92
d. Randomness factor was taken as 0.90
e. Absorption coefficient as 1
f. Randomness reduction was taken as 0.74
g. Number of parameters are 5.

Trial runs were conducted to select the aforementioned control parameters. Furthermore, the solution for the optimal solution was revealed to be in close proximity to the obtained results for the problem under consideration. The implementation of the firefly algorithm has been discussed as follows.

A random initial position of fireflies is generated within the range of independent variable. The obtained values for the parameters are inserted in the objective function and correspondingly the objective function value is obtained using the firefly position.

Pop. No.	x_1	x_2	x_3	x_4	x_5	Intensity
1	1.91711989	3.62982524	14.4748639	37.6329876	281.220458	0.965
2	2.00940628	3.53936665	27.5754979	41.0034467	115.479668	0.825
3	1.69568438	3.4707039	11.7649697	35.0535516	167.81794	0.86
4	1.71797795	4.06898295	19.4090794	54.0332583	216.234609	0.974
5	2.5809249	3.92656965	26.276722	50.0370943	261.075198	1.003
6	2.4819249	3.812656965	26.276722	50.1370943	261.075098	1.004
7	1.87966752	2.93305292	13.69529	37.88839	206.266739	0.557
8	1.80092234	3.47586476	13.7430794	41.997409	145.573077	0.809
9	2.32948492	3.52028328	13.4089559	38.0379223	241.997996	0.745
10	1.75278438	3.36077567	23.9398424	44.9455247	129.837622	0.886
11	1.28094995	3.57085557	14.3863805	51.2840621	231.734347	0.893
12	2.51538943	3.19595392	15.9730983	47.7684869	226.89762	0.662
13	1.4899535	4.06539399	25.4778039	48.87904	145.97236	1.168
14	1.80409629	3.37899883	20.0340029	47.8925055	136.556704	0.855
15	1.95063583	3.93422396	28.2993972	49.6975456	133.380603	1.058
16	1.96866779	3.57297792	11.1680692	52.2979253	129.932529	0.663
17	0.93662903	3.96947055	18.8460497	47.6492087	140.65959	1.072
18	0.77435786	4.48622959	21.5560959	38.7229386	290.99289	0.964
19	1.05996993	3.64560992	21.4024543	46.5772409	103.292779	0.99
20	2.34604788	4.4072486	26.5874424	38.3824342	291.60398	1.087

The sorting process during the initialization phase has been depicted in Table 12.2.

Results for the first iteration by employing firefly algorithm for electrochemical micro machining process have been tabulated Table 12.3.

Optimizing Engineering Problems

TABLE 12.2
Initialization Sorting

Pop. No.	x_1	x_2	x_3	x_4	x_5	Distance
13	1.4899535	4.06539399	25.4778039	48.87904	145.97236	0
20	2.34604788	4.4072486	26.5874424	38.3824342	291.60398	167.396
17	0.93662903	3.96947055	18.8460497	47.6492087	140.65959	123.674
15	1.95063583	3.93422396	28.2993972	49.6975456	133.380603	68.439
7	1.87966752	2.93305292	13.69529	37.88839	206.266739	88.299
19	1.05996993	3.64560992	21.4024543	46.5772409	103.292779	27.593
4	1.71797795	4.06898295	19.4090794	54.0332583	216.234609	76.074
18	0.77435786	4.48622959	21.5560959	38.7229386	290.99289	136.937
1	1.91711989	3.62982524	14.4748639	37.6329876	281.220458	38.436
11	1.28094995	3.57085557	14.3863805	51.2840621	231.734347	162.871
10	1.75278438	3.36077567	23.9398424	44.9455247	129.837622	52.548
3	1.69568438	3.4707039	11.7649697	35.0535516	167.81794	56.487
14	1.80409629	3.37899883	20.0340029	47.8925055	136.556704	137.373
2	2.00940628	3.53936665	27.5754979	41.0034467	115.479668	146.349
8	1.80092234	3.47586476	13.7430794	41.997409	145.573077	150.084
5	2.5809249	3.92656965	26.276722	50.0370943	261.075198	163.043
9	2.32948492	3.52028328	13.4089559	38.0379223	241.997996	142.099
16	1.96866779	3.57297792	11.1680692	52.2979253	129.932529	13.322
12	2.51538943	3.19595392	15.9730983	47.7684869	226.89762	179.490
6	2.4819249	3.812656965	26.276722	50.1370943	261.075098	16.940

TABLE 12.3
Results Obtained for First Iteration Using Firefly Algorithm

Pop. No.	x_1	x_2	x_3	x_4	x_5	Intensity
1	0.69994805	3.37624993	10.63393997	49.99244246	108.9646593	0.859
2	0.65367987	3.153906959	25.82663683	43.86909246	297.7569335	0.722
3	1.58936204	4.346877397	29.56237092	36.070011206	295.3294877	1.233
4	2.29989875	3.994005925	19.957373	36.86754799	183.196705	0.879
5	2.59509399	3.49685657	26.7903442	50.96972926	229.9520909	0.863
6	0.60989633	3.929388827	12.73808482	49.22275857	114.6995705	1.007
7	1.69632593	4.499773677	26.29677953	36.75782208	142.6934994	1.265
8	2.33379746	2.959346858	28.62583472	47.25244769	108.93087795	0.532
9	2.42938905	2.999293679	26.76526259	49.4970422	108.9308795	0.532
10	1.9588343	3.569767633	25.19677953	45.14513033	184.9829737	0.92
11	1.49279485	3.775523099	14.58834668	51.47386282	231.9339539	0.938
12	2.50548973	3.49493733	16.26950839	47.96809353	226.99732173	0.669
13	1.69968233	4.355799862	25.6775097	48.97884772	147.0709679	1.187
14	1.90890009	3.586688964	20.33380966	47.99222202	136.6564098	0.868
15	2.24945672	4.226785953	28.49911389	49.89725229	133.5803093	1.043

(*Continued*)

TABLE 12.3 (*Continued*)
Results Obtained for First Iteration Using Firefly Algorithm

Pop. No.	x_1	x_2	x_3	x_4	x_5	Intensity
16	2.06960644	3.77660877	11.367744	52.49773093	130.232236	0.606
17	1.14609573	4.16947055	18.9458533	38.90385788	140.8592973	1.106
18	0.97435786	4.50622959	21.60205996	38.90385789	291.192598	1.0454
19	1.35952908	3.847474984	21.60205996	46.66694869	103.3824855	1.040
20	2.43824845	4.495807799	26.687049	38.56999963	291.8136878	1.056

The sorting process during the first iteration phase has been depicted in Table 12.4.

TABLE 12.4
First Iteration Sorting

Pop. No.	x_1	x_2	x_3	x_4	x_5	Intensity
13	1.69968233	4.355799862	25.6775097	48.97884772	147.0709679	1.187
7	1.69632593	4.499773677	26.29677953	36.75782208	142.6934994	1.265
3	1.58936204	4.346877397	29.56237092	36.070011206	295.3294877	1.233
17	1.14609573	4.16947055	18.9458533	38.90385788	140.8592973	1.106
20	2.43824845	4.495807799	26.687049	38.56999963	291.8136878	1.056
19	1.35952908	3.847474984	21.60205996	46.66694869	103.3824855	1.040
18	0.97435786	4.50622959	21.60205996	38.90385789	291.192598	1.0454
15	2.24945672	4.226785953	28.49911389	49.89725229	133.5803093	1.043
6	0.60989633	3.929388827	12.73808482	49.22275857	114.6995705	1.007
5	2.59509399	3.49685657	26.7903442	50.96972926	229.9520909	0.863
11	1.49279485	3.775523099	14.58834668	51.47386282	231.9339539	0.938
10	1.9588343	3.569767633	25.19677953	45.14513033	184.9829737	0.92
4	2.29989875	3.994005925	19.957373	36.86754799	183.196705	0.879
14	1.90890009	3.586688964	20.33380966	47.99222202	136.6564098	0.868
1	0.69994805	3.37624993	10.63393997	49.99244246	108.9646593	0.859
9	2.42938905	2.999293679	26.76526259	49.4970422	108.9308795	0.532
2	0.65367987	3.153906959	25.82663683	43.86909246	297.7569335	0.722
12	2.50548973	3.49493733	16.26950839	47.96809353	226.99732173	0.669
16	2.06960644	3.77660877	11.367744	52.49773093	130.232236	0.606
8	2.33379746	2.959346858	28.62583472	47.25244769	108.93087795	0.532

12.4 CONCLUSION

Present chapter illustrated the applicability of firefly algorithm to a non-traditional machining process. The effectiveness was demonstrated for electrochemical micro machining process and the results obtained showed that a near optimal solution was reached using the firefly algorithm.

13 Engineering Problem Optimized Using Cuckoo Search Algorithm

13.1 INTRODUCTION

Machining is part of the manufacturing of all the metal parts. Modern machining and conventional machining are the two types of machining processes. On the one hand, conventional machining process removes the workpiece material in the form of chips such as milling, drilling and turning. Abrasive water jet machining, chemical machining, photochemical machining, ultrasonic machining are the examples of modern machining processes. One of the recent modern machining methods is abrasive water jet machining process (Yusup et al., 2014) that has been employed for machining of hard materials such as intricate profiles. Some of the advantages of the abrasive water jet machining process are: higher flexibility, high versatility and minimized thermal distortion. Furthermore, small cutting force is another probable advantage of this process. Action of high speed water jet mixed with abrasive particles results in the material removal mechanism in case of abrasive water jet machining process. Because of the umpteen advantages, the abrasive jet machining process has been used widely by numerous manufacturing industries.

Surface roughness has been revealed to be one of the critical measure for machining performance (Çaydas and Ekici, 2012; Moola et al., 2012). Quality of product can be well evaluated through the surface roughness parameter and ultimately effects the cost of the product. Also, for the desired functional performance of the product, a required surface quality is of utmost importance. However, several parameters such as machining parameters, cutting tool properties, properties of the workpiece and the cutting phenomenon affect the surface roughness value (Adnan et al., 2013).

Also, the proper selection of the machining parameters is critical for the successful machining process. Selection needs to be done under the consideration of cost and quality factors. The major issue lies in accurately obtaining the surface roughness value for the enhanced machining performance. It is the operator who needs to define the cutting parameters. As such several trial runs need to be performed resulting in high cost and wastage of material. The operators find it difficult to identify the optimum points of experimentation (Xu et al., 2013).

This has interested researchers to carry out research in the domain of evolutionary optimization techniques for solving the optimization problems in the domain of non-traditional machining process. Owing to the importance of optimization and as a subject of present chapter, the cuckoo search algorithm has been employed to minimize the surface roughness value for abrasive water jet machining process. Traverse speed, standoff distance, water jet pressure, abrasive flow rate and abrasive grit size have been considered as the five machining parameters.

13.2 CUCKOO SEARCH ALGORITHM

Expansion of technology has introduced various computational techniques that have been utilized to obtain improved value for improvising the production performance. The prediction of the optimized parameters in order to improve the machining performance can be made through the utilization of evolutionary techniques. Cuckoo search is one of the evolutionary optimization techniques that steps from the life of a bird family and exists in two forms: eggs and mature cuckoo. The algorithm is based on the manner in which the cuckoo lay their egg and hence the breeding nature of the cuckoo itself. The eggs laid by the cuckoo bird in the host nest can be thrown away from the nest if the host bird identifies them as alien eggs. In another situation, the host bird can simply abandon the nest and build a new nest elsewhere.

The steps involved with the cuckoo search algorithm have been discussed next.

Step 1: Initialization is done for the cuckoo habitat with some points on the profit function i.e.,

$$\text{Habitat} = x_1, x_2, \dots, x_{N\text{var}} \tag{13.1}$$

$$\text{Profit} = f_p(\text{habitat}) = f_p(x_1, x_2, \dots, x_{N\text{var}}) \tag{13.2}$$

$$\text{Profit} = -\text{Cost}(\text{habitat})$$
$$= -f_c(x_1, x_2, \dots, x_{N\text{var}}) \tag{13.3}$$

Step 2: Some of the eggs are dedicated to each cuckoo.

Step 3: Define the value of egg laying radius (ELR) for each cuckoo. This is done using Equation (13.4):

$$\text{ELR} = \frac{\alpha \times \text{Number of current cuckoo's eggs}}{\text{Total number of eggs} \times (\text{var}_{\text{hi}} - \text{var}_{\text{low}})} \tag{13.4}$$

where, α is some integer.

Step 4: Cuckoos are allowed to lay the eggs inside their corresponding ELR.

Step 5: Host birds kill the eggs that they recognize as alien eggs.

Step 6: If the eggs are not identified, they hatch and grow.

Step 7: Habitat for each of the newly grown cuckoo is evaluated.

Step 8: Number of cuckoos in the environment is limited by killing of the cuckoo birds that live in the worst habitat.

Step 9: Cuckoos are then clustered and best group is then identified. Goal habitat is then identified.

Step 10: The cuckoos are allowed to migrate toward the identified goal habitat.
Step 11: If the termination criteria is met, then the algorithm is stopped. If not, then *Step 2* is followed.

13.3 APPLICATION OF CUCKOO SEARCH ALGORITHM TO ABRASIVE WATER JET MACHINING

The experiments were first conducted. An alloy was considered for this purpose, and an abrasive water jet cutting machine was employed. The experiments were considered with some critical parameters: traverse speed (V), pressure of the water jet (P), abrasive grit size (d), standoff distance (h) and abrasive flow rate. The different levels of the parameters considered for carrying out the experiments have been depicted in Table 13.1.

As observed from the parameters and their levels, a total of 3^5 i.c., 243 number of experiments are required to be conducted. However, in order to save the experimental time as well as the cost, Taguchi's orthogonal array was used. As such, L_{27} orthogonal array was found to be suitable for creating design of experiments. The experiments and the obtained values of surface roughness have been tabulated in Table 13.2.

The mathematical model for surface roughness was predicted using multi-linear stepwise regression analysis. The regression model, in general, can be expressed using Equation (13.5).

$$R_a = \beta_0 + \sum_{i=1}^{k} \beta_i X_i + \sum_{i=1}^{k} \beta_{ii} X_i^2 + \sum \sum_{i<1}^{k} \beta_{ij} X_i X_j + \varepsilon_i \tag{13.5}$$

where, k represents the number of factors under consideration.

Equation (13.6) represents the second-order polynomial regression equation.

$$R_a = b_0 + b_1 x_1 + b_2 x_2 + b_3 x_3 + b_4 x_4 + b_5 x_5 + b_{11} x_1^2 + b_{22} x_2^2 + b_{33} x_3^2 + b_{44} x_4^2$$

$$+ b_{55} x_5^2 + b_{12} x_1 x_2 + b_{13} x_1 x_3 + b_{14} x_1 x_4 + b_{15} x_1 x_5 + b_{23} x_2 x_3 + b_{24} x_2 x_4$$

$$+ b_{25} x_2 x_5 + b_{34} x_3 x_4 + b_{35} x_3 x_5 + b_{45} x_4 x_5 \tag{13.6}$$

TABLE 13.1
Machine Settings Used in the Experiments

Parameter	Level 1	Level 2	Level 3
Traverse speed (mm/min) (x_1)	75	125	175
Water jet pressure (MPa) (x_2)	135	185	235
Standoff distance (mm) (x_3)	1.5	3	3.5
Abrasive grit size (µm) (x_4)	60	90	120
Abrasive flow rate (g/s) (x_5)	1	2	3

TABLE 13.2
Design of Experiments Using L$_{27}$ Orthogonal Array

Exp. No.	x_1	x_2	x_3	x_4	x_5	Surface Roughness (μm)
1	75	135	1.5	60	1	2.235
2	75	135	1.5	60	2	2.864
3	75	135	1.5	60	3	3.463
4	75	185	3	90	1	4.422
5	75	185	3	90	2	4.652
6	75	185	3	90	3	5.234
7	75	235	3.5	120	1	6.899
8	75	235	3.5	120	2	7.635
9	75	235	3.5	120	3	9.234
10	125	135	3	120	1	3.686
11	125	135	3	120	2	4.568
12	125	135	3	120	3	5.739
13	125	185	3.5	60	1	7.121
14	125	185	3.5	60	2	7.646
15	125	185	3.5	60	3	7.994
16	125	235	1.5	90	1	8.232
17	125	235	1.5	90	2	8.423
18	125	235	1.5	90	3	9.274
19	175	135	3.5	90	1	4.439
20	175	135	3.5	90	2	5.231
21	175	135	3.5	90	3	5.963
22	175	185	3	120	1	6.254
23	175	185	3	120	2	6.832
24	175	185	3	120	3	7.891
25	175	235	3	60	1	8.991
26	175	235	3	60	2	9.231
27	175	235	3	60	3	10.146

The cuckoo algorithm was run for different initial eggs, and then the surface roughness value was obtained. The values obtained have been tabulated in Table 13.3.

The values obtained were compared with those of the experimental results. It was revealed that the value of surface roughness decreased with the increasing number of initial eggs.

TABLE 13.3
Predicted Values of Surface Roughness Obtained Using Cuckoo Algorithm

Exp. No.	x_1	x_2	x_3	x_4	x_5	Initial Eggs = 10	Initial Eggs = 20	Initial Eggs = 30
1	75	135	1.5	60	1	2.4378	2.5378	2.6378
2	75	135	1.5	60	2	2.8925	2.9925	2.9945
3	75	135	1.5	60	3	3.2994	3.2994	3.2994
4	75	185	3	90	1	3.9578	3.9578	3.9578
5	75	185	3	90	2	4.6923	4.6923	4.6923
6	75	185	3	90	3	5.4545	5.4545	5.4545
7	75	235	3.5	120	1	6.6799	6.6799	6.6799
8	75	235	3.5	120	2	6.9906	6.9906	6.9906
9	75	235	3.5	120	3	8.5289	8.5289	8.5289
10	125	135	3	120	1	3.8809	3.8809	3.8809
11	125	135	3	120	2	4.6466	4.6466	4.6466
12	125	135	3	120	3	5.457	5.457	5.457
13	125	185	3.5	60	1	6.9929	6.9929	6.9929
14	125	185	3.5	60	2	7.4292	7.4292	7.4292
15	125	185	3.5	60	3	8.4564	8.4564	8.4564
16	125	235	1.5	90	1	8.3495	8.3495	8.3495
17	125	235	1.5	90	2	8.4535	8.4535	8.4535
18	125	235	1.5	90	3	9.3636	9.3636	9.3636
19	175	135	3.5	90	1	4.439	4.439	4.439
20	175	135	3.5	90	2	4.9156	4.9156	4.9156
21	175	135	3.5	90	3	5.8912	5.8912	5.8912
22	175	185	3	120	1	5.9294	5.9294	5.9294
23	175	185	3	120	2	6.6769	6.6769	6.6769
24	175	185	3	120	3	7.6326	7.6326	7.6326
25	175	235	3	60	1	8.7189	8.7189	8.7189
26	175	235	3	60	2	8.9616	8.9616	8.9616
27	175	235	3	60	3	9.6529	9.6529	9.6529

13.4 CONCLUSION

The present chapter demonstrated the applicability of the cuckoo search algorithm for the non-traditional machining process. As such the demonstration was done for the abrasive water jet machining process, and it was revealed that the algorithm was effective in estimating the surface roughness value. Analysis on the initial eggs laid revealed that the value of surface roughness became smaller as the number of initial eggs laid increased. As such the value of the parameter under investigation reached nearer to the goal point.

References

Abbass, H. A. (2001, May). MBO: marriage in honey bees optimization-A haplometrosis polygynous swarming approach. In *Proceedings of the 2001 Congress on Evolutionary Computation (IEEE Cat. No. 01TH8546)*, Vol. 1, pp. 207–214. IEEE, Seoul.

Abdullah, S., & Alzaqebah, M. (2013). A hybrid self-adaptive bees algorithm for examination timetabling problems. *Applied Soft Computing*, 13, 3608–3620.

Abdullah, A., Deris, S., Mohamad, M. S., & Hashim, S. Z. M. (2012). A new hybrid firefly algorithm for complex and nonlinear problem. In *Distributed Computing and Artificial Intelligence*, pp. 673–680. Springer, Berlin and Heidelberg.

Abedinia, O., Amjady, N., Ghasemi, A., & Hejrati, Z. (2013). Solution of economic load dispatch problem via hybrid particle swarm optimization with time-varying acceleration coefficients and bacteria foraging algorithm techniques. *International Transactions on Electrical Energy Systems*, 23(8), 1504–1522.

Abedinia, O., Naslian, M. D., & Bekravi, M. (2014). A new stochastic search algorithm bundled honeybee mating for solving optimization problems. *Neural Computing and Applications*, 25(7–8), 1921–1939.

Ahn, C. W., & Ramakrishna, R. S. (2003). Elitism-based compact genetic algorithms. *IEEE Transactions on Evolutionary Computation*, 7(4), 367–385.

Al-Anzi, F. S., & Allahverdi, A. (2007). A self-adaptive differential evolution heuristic for two-stage assembly scheduling problem to minimize maximum lateness with setup times. *European Journal of Operational Research*, 182(1), 80–94.

Ang, M. C., Pham, D. T., & Ng, K. W. (2009, June). Minimum-time motion planning for a robot arm using the bees algorithm. In *2009 7th IEEE International Conference on Industrial Informatics*, pp. 487–492. IEEE.

Ankenbrandt, C. A. (1991). An extension to the theory of convergence and a proof of the time complexity of genetic algorithms. *Foundations of Genetic Algorithms*, 1, 53–68.

Asoh, H., & Mühlenbein, H. (1994, October). On the mean convergence time of evolutionary algorithms without selection and mutation. In *International Conference on Parallel Problem Solving from Nature*, pp. 88–97. Springer, Berlin and Heidelberg.

Azarbad, M., Ebrahimzade, A., & Izadian, V. (2011). Segmentation of infrared images and objectives detection using maximum entropy method based on the bee algorithm. *International Journal of Computer Information Systems and Industrial Management Applications*, 3, 26–33.

Bäck, T., Fogel, D. B., & Michalewicz, Z. (Eds.). (2018). *Evolutionary Computation 1: Basic Algorithms and Operators*. CRC Press, Boca Raton, FL.

Bagheri, A., Peyhani, H. M., & Akbari, M. (2014). Financial forecasting using ANFIS networks with quantum-behaved particle swarm optimization. *Expert Systems with Applications*, 41(14), 6235–6250.

Bahamish, H. A. A., Abdullah, R., & Salam, R. A. (2008, May). Protein conformational search using bees algorithm. In *2008 Second Asia International Conference on Modelling & Simulation (AMS)*, pp. 911–916. IEEE.

Bai, Q. (2010). Analysis of particle swarm optimization algorithm. *Computer and Information Science*, 3(1), 180.

Bai, J., Yang, G. K., Chen, Y. W., Hu, L. S., & Pan, C. C. (2013). A model induced max-min ant colony optimization for asymmetric traveling salesman problem. *Applied Soft Computing*, 13(3), 1365–1375.

Baluja, S. (1994). Population-based incremental learning. A method for integrating genetic search based function optimization and competitive learning (No. CMU-CS-94–163). Carnegie-Mellon University, Pittsburgh, PA, Department of Computer Science.

Barthelemy, J. F., & Haftka, R. T. (1993). Approximation concepts for optimum structural design—A review. *Structural Optimization*, 5(3), 129–144.

Baskar, N., Asokan, P., Prabhaharan, G., & Saravanan, R. (2005). Optimization of machining parameters for milling operations using non-conventional methods. *The International Journal of Advanced Manufacturing Technology*, 25(11–12), 1078–1088.

Blickle, T., & Thiele, L. (1996). A comparison of selection schemes used in evolutionary algorithms. *Evolutionary Computation*, 4(4), 361–394.

Bonabeau, E., Marco, D. D. R. D. F., Dorigo, M., & Theraulaz, G. (1999). *Swarm Intelligence: From Natural to Artificial Systems (No. 1)*. Oxford University Press, New York.

Boonyaritdachochai, P., Boonchuay, C., & Ongsakul, W. (2010, June). Optimal congestion management in electricity market using particle swarm optimization with time varying acceleration coefficients. In *AIP Conference Proceedings*, Vol. 1239, No. 1, pp. 382–387. American Institute of Physics (AIP), Melville, NY.

Brasier, A. R., Tate, J. E., & Habener, J. F. (1989). Optimized use of the firefly luciferase assay as a reporter gene in mammalian cell lines. *Biotechniques*, 7(10), 1116–1122.

Brown, C. T., Liebovitch, L. S., & Glendon, R. (2007). Lévy flights in Dobe Ju/'hoansi foraging patterns. *Human Ecology*, 35(1), 129–138.

Burke, E., Cowling, P., De Causmaecker, P., & Berghe, G. V. (2001). A memetic approach to the nurse rostering problem. *Applied Intelligence*, 15(3), 199–214.

Burke, E. K., & Smith, A. J. (2000). Hybrid evolutionary techniques for the maintenance scheduling problem. *IEEE Transaction on Power Systems*, 15, 122–128.

Carrión, M., & Arroyo, J. M. (2006). A computationally efficient mixed-integer linear formulation for the thermal unit commitment problem. *IEEE Transactions on Power Systems*, 21(3), 1371–1378.

Çaydaş, U., & Ekici, S. (2012). Support vector machines models for surface roughness prediction in CNC turning of AISI 304 austenitic stainless steel. *Journal of Intelligent Manufacturing*, 23(3), 639–650.

Cetin, M. H., Ozcelik, B., Kuram, E., & Demirbas, E. (2011). Evaluation of vegetable based cutting fluids with extreme pressure and cutting parameters in turning of AISI 304L by Taguchi method. *Journal of Cleaner Production*, 19(17–18), 2049–2056.

Chandrasekaran, K., & Simon, S. P. (2012). Network and reliability constrained unit commitment problem using binary real coded firefly algorithm. *International Journal of Electrical Power & Energy Systems*, 43(1), 921–932.

Chaturvedi, K. T., Pandit, M., & Srivastava, L. (2009). Particle swarm optimization with time varying acceleration coefficients for non-convex economic power dispatch. *International Journal of Electrical Power & Energy Systems*, 31(6), 249–257.

Chen, G. (2010, August). Simplified particle swarm optimization algorithm based on particles classification. In *2010 Sixth International Conference on Natural Computation*, Vol. 5, pp. 2701–2705. IEEE.

Chen, J., Tang, Y., Ge, R., An, Q., & Guo, X. (2013). Reliability design optimization of composite structures based on PSO together with FEA. *Chinese Journal of Aeronautics*, 26(2), 343–349.

Chen, J., Tang, Y., & Huang, X. (2013). Application of surrogate based particle swarm optimization to the reliability-based robust design of composite pressure vessels. *Acta Mechanica Solida Sinica*, 26(5), 480–490.

Chuang, L. Y., Tsai, S. W., & Yang, C. H. (2011). Chaotic catfish particle swarm optimization for solving global numerical optimization problems. *Applied Mathematics and Computation*, 217(16), 6900–6916.

Clerc, M. (1999, July). The swarm and the queen: towards a deterministic and adaptive particle swarm optimization. In *Proceedings of the 1999 Congress on Evolutionary Computation-CEC99 (Cat. No. 99TH8406*, Vol. 3, pp. 1951–1957. IEEE.

Clerc, M., & Kennedy, J. (2002). The particle swarm-explosion, stability, and convergence in a multidimensional complex space. *IEEE Transactions on Evolutionary Computation*, 6(1), 58–73.

Collins, R. J., & Jefferson, D. R. (1991). *Selection in Massively Parallel Genetic Algorithms* (pp. 249–256). Computer Science Department, University of California, Los Angeles, CA.

Colorni, A., Dorigo, M., & Maniezzo, V. (1992, December). Distributed optimization by ant colonies. In *Proceedings of the First European Conference on Artificial Life*, Vol. 142, pp. 134–142.

Costa, D. (1995). An evolutionary tabu search algorithm and the NHL scheduling problem. *INFOR: Information Systems and Operational Research*, 33(3), 161–178.

Crow, J. F., & Kimura, M. (1970). *An Introduction to Population Genetics Theory*. Harper and Row, New York.

Dai, Y., Wei, Y., Chen, J., Zhang, Y., & Ding, J. (2012, October). Seismic wavelet estimation based on adaptive chaotic embedded particle swarm optimization algorithm. In *2012 Fifth International Symposium on Computational Intelligence and Design*, Vol. 2, pp. 57–60. IEEE.

Dam, H., Quist, P., & Schreiber, M. P. (1995). Productivity, surface quality and tolerances in ultrasonic machining of ceramics. *Journal of Materials Processing Technology*, 51(1–4), 358–368.

Das, S., Maity, S., Qu, B. Y., & Suganthan, P. N. (2011). Real-parameter evolutionary multimodal optimization—a survey of the state-of-the-art. *Swarm and Evolutionary Computation*, 1(2), 71–88.

Davis, L. (1985, August). Applying adaptive algorithms to epistatic domains. In *IJCAI*, Vol. 85, pp. 162–164.

Davoodi, E., Hagh, M. T., & Zadeh, S. G. (2014). A hybrid improved quantum-behaved particle swarm optimization–simplex method (IQPSOS) to solve power system load flow problems. *Applied Soft Computing*, 21, 171–179.

De Wet, J. R., Wood, K. V., DeLuca, M., Helinski, D. R., & Subramani, S. (1987). Firefly luciferase gene: structure and expression in mammalian cells. *Molecular and Cellular Biology*, 7(2), 725–737.

Deb, K., & Goldberg, D. E. (1994). Sufficient conditions for deceptive and easy binary functions. *Annals of Mathematics and Artificial Intelligence*, 10(4), 385–408.

Deep, K., & Thakur, M. (2007). A new mutation operator for real coded genetic algorithms. *Applied Mathematics and Computation*, 193(1), 211–230.

Dereli, T., & Das, G. S. (2011). A hybrid 'bee (s) algorithm' for solving container loading problems. *Applied Soft Computing*, 11(2), 2854–2862.

Dhupal, D., Doloi, B., & Bhattacharyya, B. (2007). Optimization of process parameters of Nd: YAG laser microgrooving of Al₂TiO₅ ceramic material by response surface methodology and artificial neural network algorithm. *Proceedings of the Institution of Mechanical Engineers, Part B: Journal of Engineering Manufacture*, 221(8), 1341–1350.

Dhupal, D., Doloi, B., & Bhattacharyya, B. (2008). Pulsed Nd: YAG laser turning of microgroove on aluminum oxide ceramic (Al₂O₃). *International Journal of Machine Tools and Manufacture*, 48(2), 236–248.

Diwold, K., Beekman, M., & Middendorf, M. (2011). Honeybee optimisation–an overview and a new bee inspired optimisation scheme. In *Handbook of Swarm Intelligence*, pp. 295–327. Springer, Berlin and Heidelberg.

Dorigo, M. (1992). Optimization, learning and natural algorithms. *PhD Thesis*, Politecnico di Milano.

Dorigo, M., & Gambardella, L. M. (1997). Ant colony system: a cooperative learning approach to the traveling salesman problem. *IEEE Transactions on Evolutionary Computation*, 1(1), 53–66.

Dorigo, M., & Stützle, T. (2004). *Ant Colony Optimization*. Massachusetts Institute of Technology, Cambridge.

Dos Santos, R. P. B., Martins, C. H., & Santos, F. L. (2012). Simplified particle swarm optimization algorithm. *Acta Scientiarum Technology*, 34(1), 21–25.

Dos Santos Coelho, L., de Andrade Bernert, D. L., & Mariani, V. C. (2011, June). A chaotic firefly algorithm applied to reliability-redundancy optimization. In *2011 IEEE Congress of Evolutionary Computation (CEC)*, pp. 517–521. IEEE.

Dubey, A. K., & Yadava, V. (2008a). Experimental study of Nd: YAG laser beam machining—an overview. *Journal of Materials Processing Technology*, 195(1–3), 15–26.

Dubey, A. K., & Yadava, V. (2008b). Laser beam machining—a review. *International Journal of Machine Tools and Manufacture*, 48(6), 609–628.

Dubey, A. K., & Yadava, V. (2008c). Multi-objective optimisation of laser beam cutting process. *Optics & Laser Technology*, 40(3), 562–570.

Eiben, A. E., Van Der Hauw, J. K., & van Hemert, J. I. (1998). Graph coloring with adaptive evolutionary algorithms. *Journal of Heuristics*, 4(1), 25–46.

Eshelman, L. J., Caruana, R. A., & Schaffer, J. D. (1989). Biases in the crossover landscape. In *Proceedings of the Third International Conference on Genetic Algorithms*. George Mason University, Fairfax, Virginia, USA.

Falcon, R., Almeida, M., & Nayak, A. (2011, June). Fault identification with binary adaptive fireflies in parallel and distributed systems. In *2011 IEEE Congress of Evolutionary Computation (CEC)*, pp. 1359–1366. IEEE.

Farahani, S. M., Abshouri, A. A., Nasiri, B., & Meybodi, M. R. (2011). A Gaussian firefly algorithm. *International Journal of Machine Learning and Computing*, 1(5), 448.

Fister, I., Fister Jr, I., Yang, X. S., & Brest, J. (2013). A comprehensive review of firefly algorithms. *Swarm and Evolutionary Computation*, 13, 34–46.

Fister, I., Mernik, M., & Filipič, B. (2013). Graph 3-coloring with a hybrid self-adaptive evolutionary algorithm. *Computational Optimization and Applications*, 54(3), 741–770.

Fister Jr, I., Yang, X. S., Fister, I., & Brest, J. (2012). Memetic firefly algorithm for combinatorial optimization. *arXiv preprint arXiv*, 1204, 5165.

Fogel, D. B., & Atmar, J. W. (1990). Comparing genetic operators with Gaussian mutations in simulated evolutionary processes using linear systems. *Biological Cybernetics*, 63(2), 111–114.

Fonseca, C. M., & Fleming, P. J. (1995). An overview of evolutionary algorithms in multiobjective optimization. *Evolutionary Computation*, 3(1), 1–16.

Forouzan, A. B. (2007). *Data Communications & Networking (sie)*. Tata McGraw-Hill Education.

Fratila, D., & Caizar, C. (2011). Application of Taguchi method to selection of optimal lubrication and cutting conditions in face milling of $AlMg_3$. *Journal of Cleaner Production*, 19(6–7), 640–645.

Gambardella, L. M., & Dorigo, M. (1996, May). Solving symmetric and asymmetric TSPs by ant colonies. In *Proceedings of IEEE International Conference on Evolutionary Computation*, pp. 622–627. IEEE.

Gandomi, A. H., Yang, X. S., & Alavi, A. H. (2013). Cuckoo search algorithm: a metaheuristic approach to solve structural optimization problems. *Engineering with Computers*, 29(1), 17–35.

Gandomi, A. H., Yang, X. S., Talatahari, S., & Alavi, A. H. (2013). Firefly algorithm with chaos. *Communications in Nonlinear Science and Numerical Simulation*, 18(1), 89–98.

Garnier, S., Gautrais, J., & Theraulaz, G. (2007). The biological principles of swarm intelligence. *Swarm Intelligence*, 1(1), 3–31.

Genlin, J. (2004). Survey on genetic algorithm. *Computer Applications and Software*, 2, 69–73.

Ghanbarzadeh, A. (2007). The bees algorithm: a novel optimisation tool. *PhD Thesis*, Cardiff University.

Gholizadeh, S., & Moghadas, R. K. (2014). Performance-based optimum design of steel frames by an improved quantum particle swarm optimization. *Advances in Structural Engineering*, 17(2), 143–156.

Goldberg, D. E. (1990). A note on Boltzmann tournament selection for genetic algorithms and population-oriented simulated annealing. *Complex Systems*, 4(4), 445–460.

Goldberg, D. E., & Holland, J. H. (1988). Genetic algorithms and machine learning. *Machine Learning*, 3(2), 95–99.

Goldberg, D. E., Deb, K., & Horn, J. (1992, April). Massive multimodality, deception, and genetic algorithms. In *PPSN*, Vol. 2.

Goldberg, D. E., Korb, B., & Deb, K. (1989). Messy genetic algorithms: motivation, analysis, and first results. *Complex Systems*, 3(5), 493–530.

Goldberg, D. E., Sastry, K., & Latoza, T. (2001, July). On the supply of building blocks. In *Proceedings of the 3rd Annual Conference on Genetic and Evolutionary Computation*, pp. 336–342. Morgan Kaufmann Publishers Inc., Burlington, MA.

Goldberg, D. E., & Segrest, P. (1987, July). Finite Markov chain analysis of genetic algorithms. In *Proceedings of the Second International Conference on Genetic Algorithms*, Vol. 1, p. 1.

Grefenstette, J. J. (1989). How genetic algorithms work: a critical look at implicit parallelism. In *Genetic Algorithm and Their Applications: Proceedings of Third International Conference of Genetic Algorithm*.

Grefenstette, J. J., & Fitzpatrick, J. M. (1985, July). Genetic search with approximate function evaluations. In *Proceedings of an International Conference on Genetic Algorithms and Their Applications*, pp. 112–120.

Hanafi, I., Khamlichi, A., Cabrera, F. M., Almansa, E., & Jabbouri, A. (2012). Optimization of cutting conditions for sustainable machining of PEEK-CF30 using TiN tools. *Journal of Cleaner Production*, 33, 1–9.

Harik, G., Cantú-Paz, E., Goldberg, D. E., & Miller, B. L. (1999). The gambler's ruin problem, genetic algorithms, and the sizing of populations. *Evolutionary Computation*, 7(3), 231–253.

Hassanzadeh, T., & Meybodi, M. R. (2012, May). A new hybrid algorithm based on firefly algorithm and cellular learning automata. In *20th Iranian Conference on Electrical Engineering (ICEE2012)*, pp. 628–633. IEEE.

Hertz, A., & de Werra, D. (1987). Using tabu search techniques for graph coloring. *Computing*, 39(4), 345–351.

Hinterding, R. (1995, November). Gaussian mutation and self-adaption for numeric genetic algorithms. In *Proceedings of 1995 IEEE International Conference on Evolutionary Computation*, Vol. 1, p. 384. IEEE.

Holland, J. H. (1992). Genetic algorithms. *Scientific American*, 267(1), 66–73.

Hu, X. B., & Di Paolo, E. (2009). An efficient genetic algorithm with uniform crossover for the multi-objective airport gate assignment problem. In *Multi-Objective Memetic Algorithms*, pp. 71–89. Springer, Berlin and Heidelberg.

Hussein, W. A., Sahran, S., & Abdullah, S. N. H. S. (2016). A fast scheme for multilevel thresholding based on a modified bees algorithm. *Knowledge-Based Systems*, 101, 114–134.

Husselmann, A. V., & Hawick, K. A. (2012, 16–19 July). Parallel parametric optimisation with firefly algorithms on graphical processing units. In *Proceedings of International Conference on Genetic and Evolutionary Methods (GEM12)*. Number CSTN-141, pp. 77–83, CSREA, Las Vegas, NV.

Hutter, M. (2002, May). Fitness uniform selection to preserve genetic diversity. In *Proceedings of the 2002 Congress on Evolutionary Computation. CEC'02 (Cat. No. 02TH8600)*, Vol. 1, pp. 783–788. IEEE.

Imanguliyev, A. (2013). Enhancements for the bees algorithm. *Doctoral dissertation*, Cardiff University.

Ishibuchi, H., & Yamamoto, T. (2004). Fuzzy rule selection by multi-objective genetic local search algorithms and rule evaluation measures in data mining. *Fuzzy Sets and Systems*, 141(1), 59–88.

Ismail, B. M., Reddy, B. E., & Reddy, T. B. (2018). Cuckoo inspired fast search algorithm for fractal image encoding. *Journal of King Saud University-Computer and Information Sciences*, 30(4), 462–469.

Jadoun, R. S., Kumar, P., Mishra, B. K., & Mehta, R. C. S. (2006). Manufacturing process optimisation for tool wear rate in ultrasonic drilling of engineering ceramics using the Taguchi method. *International Journal of Machining and Machinability of Materials*, 1(1), 94–114.

Jalali, M. R., Afshar, A., & Marino, M. A. (2007). Multi-colony ant algorithm for continuous multi-reservoir operation optimization problem. *Water Resources Management*, 21(9), 1429–1447.

Jamalipour, M., Sayareh, R., Gharib, M., Khoshahval, F., & Karimi, M. R. (2013). Quantum behaved particle swarm optimization with differential mutation operator applied to WWER-1000 in-core fuel management optimization. *Annals of Nuclear Energy*, 54, 134–140.

Jau, Y. M., Su, K. L., Wu, C. J., & Jeng, J. T. (2013). Modified quantum-behaved particle swarm optimization for parameters estimation of generalized nonlinear multi-regressions model based on Choquet integral with outliers. *Applied Mathematics and Computation*, 221, 282–295.

Jayswal, S. C., Jain, V. K., & Dixit, P. (2005). Modeling and simulation of magnetic abrasive finishing process. *The International Journal of Advanced Manufacturing Technology*, 26(5–6), 477–490.

Jia, P., Tian, F., Fan, S., He, Q., Feng, J. X., & Yang, S. (2014). A novel sensor array and classifier optimization method of electronic nose based on enhanced quantum-behaved particle swarm optimization. *Sensor Review*, 34(3), 304–311.

Kalatehjari, R., Rashid, A., Safuan, A., Ali, N., & Hajihassani, M. (2014). The contribution of particle swarm optimization to three-dimensional slope stability analysis. *The Scientific World Journal*, 2014, 1–12.

Karaboga, D., & Akay, B. (2009a). A comparative study of artificial bee colony algorithm. *Applied Mathematics and Computation*, 214(1), 108–132.

Karaboga, D., & Akay, B. (2009b). A survey: algorithms simulating bee swarm intelligence. *Artificial Intelligence Review*, 31(1–4), 61–85.

Karaboga, D., & Basturk, B. (2007a). A powerful and efficient algorithm for numerical function optimization: artificial bee colony (ABC) algorithm. *Journal of Global Optimization*, 39(3), 459–471.

Karaboga, D., & Basturk, B. (2007b, June). Artificial bee colony (ABC) optimization algorithm for solving constrained optimization problems. In *International Fuzzy Systems Association World Congress*, pp. 789–798. Springer, Berlin and Heidelberg.

Karaboga, D., & Basturk, B. (2008). On the performance of artificial bee colony (ABC) algorithm. *Applied Soft Computing*, 8(1), 687–697.

Karaboga, D., Gorkemli, B., Ozturk, C., & Karaboga, N. (2014). A comprehensive survey: artificial bee colony (ABC) algorithm and applications. *Artificial Intelligence Review*, 42(1), 21–57.

Kaveh, A., & Bakhshpoori, T. (2013). Optimum design of steel frames using cuckoo search algorithm with Lévy flights. *The Structural Design of Tall and Special Buildings*, 22(13), 1023–1036.

Kennedy, R., & Eberhart, J. (1995, November). Particle swarm optimization. In *Proceedings of IEEE International Conference on Neural Networks IV*, Vol. 1000, p. 33.

Kitak, P., Glotic, A., & Ticar, I. (2014). Heat transfer coefficients determination of numerical model by using particle swarm optimization. *IEEE Transactions on Magnetics*, 50(2), 933–936.

Kothari, V., Anuradha, J., Shah, S., & Mittal, P. (2011, December). A survey on particle swarm optimization in feature selection. In *International Conference on Computing and Communication Systems*, pp. 192–201. Springer, Berlin and Heidelberg.

Krishnanand, K. N., & Ghose, D. (2005, June). Detection of multiple source locations using a glowworm metaphor with applications to collective robotics. In *Proceedings 2005 IEEE Swarm Intelligence Symposium, 2005. SIS 2005*, pp. 84–91. IEEE.

Kumar, R., & Jyotishree (2012). Blending roulette wheel selection & rank selection in genetic algorithms. *International Journal of Machine Learning and Computing*, 2(4), 365–370.

Kuram, E., Ozcelik, B., Huseyin Cetin, M., Demirbas, E., & Askin, S. (2013). Effects of blended vegetable-based cutting fluids with extreme pressure on tool wear and force components in turning of Al 7075-T6. *Lubrication Science*, 25(1), 39–52.

Lake, J. J., Duwel, A. E., & Candler, R. N. (2013). Particle swarm optimization for design of slotted MEMS resonators with low thermoelastic dissipation. *Journal of Microelectromechanical Systems*, 23(2), 364–371.

Lara, C., Flores, J. J., & Calderón, F. (2008, October). Solving a school timetabling problem using a bee algorithm. In *Mexican International Conference on Artificial Intelligence*, pp. 664–674. Springer, Berlin and Heidelberg.

Lazrag, T., Kacem, M., Dubujet, P., Sghaier, J., & Bellagi, A. (2013). Determination of unsaturated hydraulic properties using drainage gravity test and particle swarm optimization algorithm. *Journal of Porous Media*, 16(11), 1025–1034.

Le Hoang, S. (2014). Optimizing municipal solid waste collection using chaotic particle swarm optimization in GIS based environments: a case study at Danang city, Vietnam. *Expert Systems with Applications*, 41(18), 8062–8074.

Lee, C. H., Shih, K. S., Hsu, C. C., & Cho, T. (2014). Simulation-based particle swarm optimization and mechanical validation of screw position and number for the fixation stability of a femoral locking compression plate. *Medical Engineering & Physics*, 36(1), 57–64.

Li, C., Zhou, J., Kou, P., & Xiao, J. (2012). A novel chaotic particle swarm optimization based fuzzy clustering algorithm. *Neurocomputing*, 83, 98–109.

Lien, L. C., & Cheng, M. Y. (2012). A hybrid swarm intelligence based particle-bee algorithm for construction site layout optimization. *Expert Systems with Applications*, 39(10), 9642–9650.

Lin, C. L., Lin, J. L., & Ko, T. C. (2002). Optimisation of the EDM process based on the orthogonal array with fuzzy logic and grey relational analysis method. *The International Journal of Advanced Manufacturing Technology*, 19(4), 271–277.

Liu, X., & Qi, D. (2016, August). Camera calibration based on self-adaptive cuckoo search algorithm. In *2016 8th International Conference on Intelligent Human-Machine Systems and Cybernetics (IHMSC)*, Vol. 2, pp. 95–98. IEEE.

Louis, S. J., & Rawlins, G. J. (1991, July). Designer genetic algorithms: genetic algorithms in structure design. In *ICGA*, pp. 53–60.

Lu, W. Z., & Xue, Y. (2014). Prediction of particulate matter at street level using artificial neural networks coupling with chaotic particle swarm optimization algorithm. *Building and Environment*, 78, 111–117.

Luthra, J., & Pal, S. K. (2011, December). A hybrid firefly algorithm using genetic operators for the cryptanalysis of a monoalphabetic substitution cipher. In *2011 World Congress on Information and Communication Technologies*, pp. 202–206. IEEE.

Madić, M., Radovanović, M., Trajanović, M., & Manić, M. (2015). Multi objective optimization of laser cutting using cuckoo search algorithm. *Journal of Engineering Science and Technology*, 10(3), 353–363.

Majumder, A. (2013). Process parameter optimization during EDM of AISI 316 LN stainless steel by using fuzzy based multi-objective PSO. *Journal of Mechanical Science and Technology*, 27(7), 2143–2151.

Majumder, A., & Laha, D. (2016). A new cuckoo search algorithm for 2-machine robotic cell scheduling problem with sequence-dependent setup times. *Swarm and Evolutionary Computation*, 28, 131–143.

Mandal, D., Pal, S. K., & Saha, P. (2007). Modeling of electrical discharge machining process using back propagation neural network and multi-objective optimization using non-dominating sorting genetic algorithm-II. *Journal of Materials Processing Technology*, 186(1–3), 154–162.

Marksberry, P. W. (2007). Micro-flood (MF) technology for sustainable manufacturing operations that are coolant less and occupationally friendly. *Journal of Cleaner Production*, 15(10), 958–971.

Mastrocinque, E., Yuce, B., Lambiase, A., & Packianather, M. S. (2013). A multi-objective optimization for supply chain network using the bees algorithm. *International Journal of Engineering Business Management*, 5, 38.

Mauldin, M. L. (1984, August). Maintaining diversity in genetic search. In *AAAI*, pp. 247–250.

Meijer, J. (2004). Laser beam machining (LBM), state of the art and new opportunities. *Journal of Materials Processing Technology*, 149(1–3), 2–17.

Miller, B. L., & Goldberg, D. E. (1995). Genetic algorithms, tournament selection, and the effects of noise. *Complex Systems*, 9(3), 193–212.

Ming, L., Hai, H., Aimin, Z., Yingde, S., Zhao, L., & Xingguo, Z. (2012). Modeling of mechanical properties of as-cast Mg-Li-Al alloys based on PSO-BP algorithm. *China Foundry*, 9(2), 119–124.

Mohamad, A., Zain, A. M., Bazin, N. E. N., & Udin, A. (2015). A process prediction model based on cuckoo algorithm for abrasive waterjet machining. *Journal of Intelligent Manufacturing*, 26(6), 1247–1252.

Mohammadi-Ivatloo, B., Moradi-Dalvand, M., & Rabiee, A. (2013). Combined heat and power economic dispatch problem solution using particle swarm optimization with time varying acceleration coefficients. *Electric Power Systems Research*, 95, 9–18.

Mohammadi-Ivatloo, B., Rabiee, A., Soroudi, A., & Ehsan, M. (2012). Iteration PSO with time varying acceleration coefficients for solving non-convex economic dispatch problems. *International Journal of Electrical Power & Energy Systems*, 42(1), 508–516.

Mohan, B. C., & Baskaran, R. (2012). A survey: ant colony optimization based recent research and implementation on several engineering domain. *Expert Systems with Applications*, 39(4), 4618–4627.

Mohan, S. C., Maiti, D. K., & Maity, D. (2013). Structural damage assessment using FRF employing particle swarm optimization. *Applied Mathematics and Computation*, 219(20), 10387–10400.

Moola, M. R., Gorin, A., & Hossein, K. A. (2012). Optimization of various cutting parameters on the surface roughness of the machinable glass ceramic with two flute square end mills of micro grain solid carbide. *International Journal of Precision Engineering and Manufacturing*, 13(9), 1549–1554.

Moradi, S., Fatahi, L., & Razi, P. (2010). Finite element model updating using bees algorithm. *Structural and Multidisciplinary Optimization*, 42(2), 283–291.

Mukherjee, R., Chakraborty, S., & Samanta, S. (2012). Selection of wire electrical discharge machining process parameters using non-traditional optimization algorithms. *Applied Soft Computing*, 12(8), 2506–2516.

Müller, S., Airaghi, S., Marchetto, J., & Koumoutsakos, P. (2000). Optimization algorithms based on a model of bacterial chemotaxis. In *Proceedings of 6th International Conference Simulation of Adaptive Behavior: From Animals to Animats*, SAB 2000 Proc. Suppl.

Munda, J., & Bhattacharyya, B. (2008). Investigation into electrochemical micromachining (EMM) through response surface methodology based approach. *The International Journal of Advanced Manufacturing Technology*, 35(7–8), 821–832.

Munetomo, M., & Goldberg, D. E. (1999). Linkage identification by non-monotonicity detection for overlapping functions. *Evolutionary Computation*, 7(4), 377–398.

Nandy, S., Sarkar, P. P., & Das, A. (2012). Analysis of a nature inspired firefly algorithm based back-propagation neural network training. *arXiv preprint arXiv*, 1206, 5360.

Nebti, S., & Boukerram, A. (2010, July). Handwritten digits recognition based on swarm optimization methods. In *International Conference on Networked Digital Technologies*, pp. 45–54. Springer, Berlin and Heidelberg.

Neubauer, A. (1997, April). A theoretical analysis of the non-uniform mutation operator for the modified genetic algorithm. In *Proceedings of 1997 IEEE International Conference on Evolutionary Computation (ICEC'97)*, pp. 93–96. IEEE.

Nguyen, K. P., & Fujita, G. (2016, November). Optimal power flow using self-learning cuckoo search algorithm. In *2016 IEEE International Conference on Power System Technology (POWERCON)*, pp. 1–6. IEEE.

Nguyen, K., Nguyen, P., & Tran, N. (2012). A hybrid algorithm of harmony search and bees algorithm for a university course timetabling problem. *International Journal of Computer Science Issues (IJCSI)*, 9(1), 12.

Noghrehabadi, A., Ghalambaz, M., & Vosough, A. (2011). A hybrid power series—cuckoo search optimization algorithm to electrostatic deflection of micro fixed-fixed actuators. *International Journal of Multidisciplinary Sciences and Engineering*, 2(4), 22–26.

Oosthuizen, G. D. (1987). Supergran: a connectionist approach to learning, integrating genetic algorithms and graph induction. In *Genetic Algorithms and Their Applications: Proceedings of the Second International Conference on Genetic Algorithms: July 28–31, 1987 at the Massachusetts Institute of Technology*, Cambridge, MA. Lawrence Erlhaum Associates, Hillsdale, NJ.

Otri, S. (2011). Improving the bees algorithm for complex optimisation problems. *Doctoral dissertation*, Cardiff University.

Ouaarab, A., Ahiod, B., & Yang, X. S. (2014). Discrete cuckoo search algorithm for the travelling salesman problem. *Neural Computing and Applications*, 24(7–8), 1659–1669.

Ozcelik, B., Kuram, E., Cetin, M. H., & Demirbas, E. (2011a). Experimental investigations of vegetable based cutting fluids with extreme pressure during turning of AISI 304L. *Tribology International*, 44(12), 1864–1871.

Ozcelik, B., Kuram, E., Demirbas, E., & Şik, E. (2011b). Optimization of surface roughness in drilling using vegetable-based cutting oils developed from sunflower oil. *Industrial Lubrication and Tribology*, 63(4), 271–276.

Packianather, M. S., & Kapoor, B. (2015, May). A wrapper-based feature selection approach using Bees Algorithm for a wood defect classification system. In *2015 10th System of Systems Engineering Conference (SoSE)*, pp. 498–503. IEEE.

Packianather, M. S., Landy, M., & Pham, D. T. (2009, June). Enhancing the speed of the bees algorithm using pheromone-based recruitment. In *2009 7th IEEE International Conference on Industrial Informatics*, pp. 789–794. IEEE.

Paechter, B., Cumming, A., Norman, M. G., & Luchian, H. (1995, August). Extensions to a memetic timetabling system. In *International Conference on the Practice and Theory of Automated Timetabling*, pp. 251–265. Springer, Berlin and Heidelberg.

Palit, S., Sinha, S. N., Molla, M. A., Khanra, A., & Kule, M. (2011, September). A cryptanalytic attack on the knapsack cryptosystem using binary firefly algorithm. In *2011 2nd International Conference on Computer and Communication Technology (ICCCT-2011)*, pp. 428–432. IEEE.

Panda, S., Mishra, D., & Biswal, B. B. (2011). Determination of optimum parameters with multi-performance characteristics in laser drilling—a grey relational analysis approach. *The International Journal of Advanced Manufacturing Technology*, 54(9–12), 957–967.

Panda, S., Sahu, B. K., & Mohanty, P. K. (2012). Design and performance analysis of PID controller for an automatic voltage regulator system using simplified particle swarm optimization. *Journal of the Franklin Institute*, 349(8), 2609–2625.

Pauline, O., Sin, H. C., Sheng, D. D. C. V., Kiong, S. C., & Meng, O. K. (2017, April). Design optimization of structural engineering problems using adaptive cuckoo search algorithm. In *2017 3rd International Conference on Control, Automation and Robotics (ICCAR)*, pp. 745–748. IEEE.

Pedersen, M. E. H., & Chipperfield, A. J. (2010). Simplifying particle swarm optimization. *Applied Soft Computing*, 10(2), 618–628.

Pelikan, M., Goldberg, D. E., & Cantú-Paz, E. (2000). Linkage learning, estimation distribution, and Bayesian networks. *Evolutionary Computation*, 8(3), 314–341.

Pham, D. T., Castellani, M., & Fahmy, A. A. (2008, July). Learning the inverse kinematics of a robot manipulator using the bees algorithm. In *2008 6th IEEE International Conference on Industrial Informatics*, pp. 493–498. IEEE.

Pham, D. T., & Darwish, A. H. (2008, July). Fuzzy selection of local search sites in the bees algorithm. In *Proceedings of the 4th Virtual International Conference on Intelligent Production Machines and Systems*, pp. 1–14.

Pham, D. T., & Darwish, H. A. (2010). Using the bees algorithm with Kalman filtering to train an artificial neural network for pattern classification. *Proceedings of the Institution of Mechanical Engineers, Part I: Journal of Systems and Control Engineering*, 224(7), 885–892.

Pham, D. T., Ghanbarzadeh, A., Koç, E., Otri, S., Rahim, S., & Zaidi, M. (2006). The bees algorithm—A novel tool for complex optimisation problems. In *Intelligent Production Machines and Systems*, pp. 454–459. Elsevier Science Ltd., Oxford.

Pham, D. T., Koc, E., Lee, J. Y., & Phrueksanant, J. (2007a, June). Using the bees algorithm to schedule jobs for a machine. In *Proceedings of Eighth International Conference on Laser Metrology, CMM and Machine Tool Performance*, pp. 430–439.

Pham, D. T., Otri, S., & Darwish, A. H. (2007b, July). Application of the bees algorithm to PCB assembly optimisation. In *Proceedings 3rd International Virtual Conference on Intelligent Production Machines and Systems (IPROMS 2007)*, pp. 511–516.

Pham, Q. T., Pham, D. T., & Castellani, M. (2012). A modified bees algorithm and a statistics-based method for tuning its parameters. *Proceedings of the Institution of Mechanical Engineers, Part I: Journal of Systems and Control Engineering*, 226(3), 287–301.

Pluhacek, M., Senkerik, R., & Zelinka, I. (2014). Particle swarm optimization algorithm driven by multichaotic number generator. *Soft Computing*, 18(4), 631–639.

Pookpunt, S., & Ongsakul, W. (2013). Optimal placement of wind turbines within wind farm using binary particle swarm optimization with time-varying acceleration coefficients. *Renewable Energy*, 55, 266–276.

Prugel-Bennett, A. (2010). Benefits of a population: five mechanisms that advantage population-based algorithms. *IEEE Transactions on Evolutionary Computation*, 14(4), 500–517.

Puertas, I., Luis, C. J., & Alvarez, L. (2004). Analysis of the influence of EDM parameters on surface quality, MRR and EW of WC–Co. *Journal of Materials Processing Technology*, 153, 1026–1032.

Rahim, E. A., & Sasahara, H. (2011). A study of the effect of palm oil as MQL lubricant on high speed drilling of titanium alloys. *Tribology International*, 44(3), 309–317.

Rajurkar, K. P., Wang, Z. Y., & Kuppattan, A. (1999). Micro removal of ceramic material (Al2O3) in the precision ultrasonic machining. *Precision Engineering*, 23(2), 73–78.

Ramasamy, S. R., Gould, J., & Workman, D. (2002). Design-of-experiments study to examine the effect of polarity on stud welding. *Welding Journal*, 81(2), 19/S–26/S.

Reynolds, A. M., Smith, A. D., Reynolds, D. R., Carreck, N. L., & Osborne, J. L. (2007). Honeybees perform optimal scale-free searching flights when attempting to locate a food source. *Journal of Experimental Biology*, 210(21), 3763–3770.

Rothlauf, F. (2006). *Representations for Genetic and Evolutionary Algorithms* (pp. 9–32). Springer, Berlin and Heidelberg.

Sadiq, A. T., & Hamad, A. G. (2010). BSA: a hybrid bees' simulated annealing algorithm to solve optimization & NP-complete problems. *Engineering and Technology Journal*, 28(2), 271–281.

Salhi, S. (2006). *Heuristic Search: The Science of Tomorrow*. Operational Research Society, Birmingham.

Sanchez, J. A., Pombo, I., Alberdi, R., Izquierdo, B., Ortega, N., Plaza, S., & Martinez-Toledano, J. (2010). Machining evaluation of a hybrid MQL-CO_2 grinding technology. *Journal of Cleaner Production*, 18(18), 1840–1849.

Sastry, K., Goldberg, D., & Kendall, G. (2005). Genetic algorithms. In *Search Methodologies*, pp. 97–125. Springer, Boston, MA.

Semenkin, E., & Semenkina, M. (2012, June). Self-configuring genetic algorithm with modified uniform crossover operator. In *International Conference in Swarm Intelligence*, pp. 414–421. Springer, Berlin and Heidelberg.

Shahzad, F., Masood, S., & Khan, N. K. (2014). Probabilistic opposition-based particle swarm optimization with velocity clamping. *Knowledge and Information Systems*, 39(3), 703–737.

Shatnawi, N., Sahran, S., & Faidzul, M. (2013). A memory-based bees algorithm: an enhancement. *Journal of Applied Science*, 13, 497–502.

Shenoy, P. D., Srinivasa, K. G., Venugopal, K. R., & Patnaik, L. M. (2005). Dynamic association rule mining using genetic algorithms. *Intelligent Data Analysis*, 9(5), 439–453.

Sivarao, T. J. S., & Ammar, S. (2010). RSM based modeling for surface roughness prediction in laser machining. *International Journal of Engineering and Technology*, 10(4), 26–32.

Smith, R. E., Dike, B. A., & Stegmann, S. A. (1995, February). Fitness inheritance in genetic algorithms. In *Proceedings of the 1995 ACM Symposium on Applied Computing*, pp. 345–350. ACM.

Sohrabpoor, H., Khanghah, S. P., Shahraki, S., & Teimouri, R. (2016). Multi-objective optimization of electrochemical machining process. *The International Journal of Advanced Manufacturing Technology*, 82(9–12), 1683–1692.

Srinivas, M., & Patnaik, L. M. (1994). Adaptive probabilities of crossover and mutation in genetic algorithms. *IEEE Transactions on Systems, Man, and Cybernetics*, 24(4), 656–667.

Storn, R., & Price, K. (1997). Differential evolution–a simple and efficient heuristic for global optimization over continuous spaces. *Journal of Global Optimization*, 11(4), 341–359.

Strehler, B. L., & Totter, J. R. (1952). Firefly luminescence in the study of energy transfer mechanisms. I. Substrate and enzyme determination. *Archives of Biochemistry and Biophysics*, 40(1), 28–41.

Stützle, T., & Dorigo, M. (1999). ACO algorithms for the traveling salesman problem. *Evolutionary Algorithms in Engineering and Computer Science*, 04, 163–183.

Su, J. C., Kao, J. Y., & Tarng, Y. S. (2004). Optimisation of the electrical discharge machining process using a GA-based neural network. *The International Journal of Advanced Manufacturing Technology*, 24(1–2), 81–90.

Subutic, M., Tuba, M., & Stanarevic, N. (2012). Parallelization of the firefly algorithm for unconstrained optimization problems. *Latest Advances in Information Science and Applications*, 22(3), 264–269.

Sun, C., Zhao, H., & Wang, Y. (2011). A comparative analysis of PSO, HPSO, and HPSO-TVAC for data clustering. *Journal of Experimental & Theoretical Artificial Intelligence*, 23(1), 51–62.

Syswerda, G. (1989). Uniform crossover in genetic algorithms. In *Proceedings of the Third International Conference on Genetic Algorithms*, pp. 2–9. Morgan Kaufmann Publishers.

Syswerda, G. (1991). A study of reproduction in generational and steady-state genetic algorithms. In *Foundations of Genetic Algorithms*, Vol. 1, pp. 94–101. Elsevier.

Tang, D., Cai, Y., Zhao, J., & Xue, Y. (2014). A quantum-behaved particle swarm optimization with memetic algorithm and memory for continuous non-linear large scale problems. *Information Sciences*, 289, 162–189.

Teimouri, R., & Baseri, H. (2014). Optimization of magnetic field assisted EDM using the continuous ACO algorithm. *Applied Soft Computing*, 14, 381–389.

Teimouri, R., & Sohrabpoor, H. (2013). Application of adaptive neuro-fuzzy inference system and cuckoo optimization algorithm for analyzing electro chemical machining process. *Frontiers of Mechanical Engineering*, 8(4), 429–442.

Tereshko, V., & Lee, T. (2002). How information-mapping patterns determine foraging behaviour of a honey bee colony. *Open Systems & Information Dynamics*, 9(02), 181–193.

Tereshko, V., & Loengarov, A. (2005). Collective decision making in honey-bee foraging dynamics. *Computing and Information Systems*, 9(3), 1.

Thoe, T. B., Aspinwall, D. K., & Wise, M. L. H. (1998). Review on ultrasonic machining. *International Journal of Machine Tools and Manufacture*, 38(4), 239–255.

Tsutsui, S., & Fujimoto, Y. (1993, June). Forking genetic algorithm with blocking and shrinking modes (fGA). In *ICGA*, pp. 206–215.

Tsutsui, S., Yamamura, M., & Higuchi, T. (1999, July). Multi-parent recombination with simplex crossover in real coded genetic algorithms. In *Proceedings of the 1st Annual Conference on Genetic and Evolutionary Computation*, Vol. 1, pp. 657–664. Morgan Kaufmann Publishers Inc.

Vastrakar, N. K., & Padhy, P. K. (2013, January). Simplified PSO PI-PD controller for unstable processes. In *2013 4th International Conference on Intelligent Systems, Modelling and Simulation*, pp. 350–354. IEEE.

Vijayaragavan, S. P., Karthik, B., Kiran Kumar, T. V. U., & Sundar Raj, M. (2013). Analysis of chaotic DC-DC converter using wavelet transform. *Middle-East Journal of Scientific Research*, 16(12), 1813–1819.

Vijayaraghavan, K., Nalini, S. K., Prakash, N. U., & Madhankumar, D. (2012). Biomimetic synthesis of silver nanoparticles by aqueous extract of Syzygium aromaticum. *Materials Letters*, 75, 33–35.

Von Frisch, K. (2014). *Bees: Their Vision, Chemical Senses, and Language*. Cornell University Press, Ithaca, NY.

Vosoughi, A. R., & Gerist, S. (2014). New hybrid FE-PSO-CGAs sensitivity base technique for damage detection of laminated composite beams. *Composite Structures*, 118, 68–73.

Watson, J. P., Rana, S., Whitley, L. D., & Howe, A. E. (1999). The impact of approximate evaluation on the performance of search algorithms for warehouse scheduling. *Journal of Scheduling*, 2(2), 79–98.

Wolpert, D. H., & Macready, W. G. (1997). No free lunch theorems for optimization. *IEEE Transactions on Evolutionary Computation*, 1(1), 67–82.

Wu, Q., Law, R., Wu, E., & Lin, J. (2013). A hybrid-forecasting model reducing Gaussian noise based on the Gaussian support vector regression machine and chaotic particle swarm optimization. *Information Sciences*, 238, 96–110.

Xu, J. H., Zhang, S. Y., Tan, J. R., & Sa, R. N. (2013). Collisionless tool orientation smoothing above blade stream surface using NURBS envelope. *Journal of Zhejiang University Science A*, 14(3), 187–197.

Yang, X. S. (2009, October). Firefly algorithms for multimodal optimization. In *International Symposium on Stochastic Algorithms*, pp. 169–178. Springer, Berlin and Heidelberg.

Yang, X. S. (2010a). A new metaheuristic bat-inspired algorithm. In *Nature Inspired Cooperative Strategies for Optimization (NICSO 2010)*, pp. 65–74. Springer, Berlin and Heidelberg.

Yang, X. S. (2010b). Firefly algorithm, Levy flights and global optimization. In *Research and Development in Intelligent Systems*, vol. XXVI, pp. 209–218. Springer, London.

Yang, X. S. (2010c). *Nature-Inspired Metaheuristic Algorithms*. Luniver Press, Frome.

Yang, X. S., & Deb, S. (2009, December). Cuckoo search via Lévy flights. In *2009 World Congress on Nature & Biologically Inspired Computing (NaBIC)*, pp. 210–214. IEEE.

Yang, X. S., & Deb, S. (2010a). Eagle strategy using Lévy walk and firefly algorithms for stochastic optimization. In *Nature Inspired Cooperative Strategies for Optimization (NICSO 2010)*, pp. 101–111. Springer, Berlin and Heidelberg.

Yang, X. S., & Deb, S. (2010b). Engineering optimisation by cuckoo search. *arXiv preprint arXiv*, 1005, 2908.

Yang, X. S., & Deb, S. (2013). Multiobjective cuckoo search for design optimization. *Computers & Operations Research*, 40(6), 1616–1624.

Yang, S. H., Srinivas, J., Mohan, S., Lee, D. M., & Balaji, S. (2009). Optimization of electric discharge machining using simulated annealing. *Journal of Materials Processing Technology*, 209(9), 4471–4475.

Yeh, W. C. (2013). New parameter-free simplified swarm optimization for artificial neural network training and its application in the prediction of time series. *IEEE Transactions on Neural Networks and Learning Systems*, 24(4), 661–665.

Yu, T. L., Goldberg, D. E., Yassine, A., & Chen, Y. P. (2003, July). Genetic algorithm design inspired by organizational theory: pilot study of a dependency structure matrix driven genetic algorithm. In *Genetic and Evolutionary Computation Conference*, pp. 1620–1621. Springer, Berlin and Heidelberg.

Yuce, B., Mastrocinque, E., Lambiase, A., Packianather, M. S., & Pham, D. T. (2014). A multi-objective supply chain optimisation using enhanced Bees Algorithm with adaptive neighbourhood search and site abandonment strategy. *Swarm and Evolutionary Computation*, 18, 71–82.

Yuce, B., Mastrocinque, E., Packianather, M. S., Lambiase, A., & Pham, D. T. (2015a). The bees algorithm and its applications. In *Handbook of Research on Artificial Intelligence Techniques and Algorithms*, pp. 122–151. IGI Global, USA.

Yuce, B., Packianather, M., Mastrocinque, E., Pham, D., & Lambiase, A. (2013). Honey bees inspired optimization method: the bees algorithm. *Insects*, 4(4), 646–662.

Yuce, B., Pham, D. T., Packianather, M. S., & Mastrocinque, E. (2015b). An enhancement to the bees algorithm with slope angle computation and hill climbing algorithm and its applications on scheduling and continuous-type optimisation problem. *Production & Manufacturing Research*, 3(1), 3–19.

Yumin, D., & Li, Z. (2014). Quantum behaved particle swarm optimization algorithm based on artificial fish swarm. *Mathematical Problems in Engineering*, 2014, 1–10.

Yusup, N., Sarkheyli, A., Zain, A. M., Hashim, S. Z. M., & Ithnin, N. (2014). Estimation of optimal machining control parameters using artificial bee colony. *Journal of Intelligent Manufacturing*, 25(6), 1463–1472.

Zeng, Y., & Sun, Y. (2014). An improved particle swarm optimization for the combined heat and power dynamic economic dispatch problem. *Electric Power Components and Systems*, 42(15), 1700–1716.

Zhang, Y., Gallipoli, D., & Augarde, C. (2013a). Parameter identification for elasto-plastic modelling of unsaturated soils from pressuremeter tests by parallel modified particle swarm optimization. *Computers and Geotechnics*, 48, 293–303.

Zhang, S., Li, J. F., & Wang, Y. W. (2012). Tool life and cutting forces in end milling Inconel 718 under dry and minimum quantity cooling lubrication cutting conditions. *Journal of Cleaner Production*, 32, 81–87.

Zhang, Q., Li, Z., Zhou, C. J., & Wei, X. P. (2013b). Bayesian network structure learning based on the chaotic particle swarm optimization algorithm. *Genetics and Molecular Research*, 12(4), 4468–4479.

Zhang, Y. D., Wang, S., & Dong, Z. (2014). Classification of Alzheimer disease based on structural magnetic resonance imaging by kernel support vector machine decision tree. *Progress in Electromagnetics Research*, 144, 171–184.

Zhang, Y., Wang, S., & Ji, G. (2015). A comprehensive survey on particle swarm optimization algorithm and its applications. *Mathematical Problems in Engineering*, 2015, 1–38.

Zhang, Y., & Wu, L. (2011). Crop classification by forward neural network with adaptive chaotic particle swarm optimization. *Sensors*, 11(5), 4721–4743.

Zhang, Y., Wu, L., & Wang, S. (2013c). UCAV path planning by fitness-scaling adaptive chaotic particle swarm optimization. *Mathematical Problems in Engineering*, 2013, 1–9.

Zhang, N., & Wunsch, D. C. (2003, May). An extended Kalman filter (EKF) approach on fuzzy system optimization problem. In *The 12th IEEE International Conference on Fuzzy Systems, 2003. FUZZ'03*, Vol. 2, pp. 1465–1470. IEEE.

Zhao, J., Lei, X., & Wu, F. X. (2016, December). Identifying protein complexes in dynamic protein-protein interaction networks based on cuckoo search algorithm. In *2016 IEEE International Conference on Bioinformatics and Biomedicine (BIBM)*, pp. 1288–1295. IEEE.

Zhu, X., & Wang, N. (2017). Cuckoo search algorithm with membrane communication mechanism for modeling overhead crane systems using RBF neural networks. *Applied Soft Computing*, 56, 458–471.

Index

Printed in the United States
by Baker & Taylor Publisher Services

Printed in the United States
by Baker & Taylor Publisher Services